WILD LILIES,

IRISES,

AND GRASSES

WILD LILIES,

IRISES,

AND GRASSES

Gardening with California Monocots

NORA HARLOW AND KRISTIN JAKOB, EDITORS

ROGER RAICHE

NEVIN SMITH

RON LUTSKO, JR.

JACOB SIGG

WAYNE RODERICK

SUZANNE SCHETTLER

CAROLINE SPILLER

LINDA HAYMAKER

JENNY FLEMING

ELIZABETH MCCLINTOCK

KRISTIN JAKOB

A PHYLLIS M. FABER BOOK

UNIVERSITY OF CALIFORNIA PRESS

Berkeley   Los Angeles   London

University of California Press
Berkeley and Los Angeles, California

University of California Press, Ltd.
London, England

Produced by Phyllis M. Faber Books
Mill Valley, California

Cover and interior design and
typesetting by Beth Hansen-Winter

Printed in Hong Kong through
Global Interprint, Santa Rosa, CA

Library of Congress Cataloguing-in-Publication Data

Wild lilies, irises, and grasses : gardening with California monocots / editors,
   Nora Harlow and Kristin Jakob ; contributors, Roger Raiche . . . [et al.].
        p. cm.
    Includes bibliographical references and index.
    ISBN 0-520-23848-6 (cloth : alk. paper) — ISBN 0-520-23849-4 (pbk. :
alk. paper)
      1. Lilies—California. 2. Irises (Plants)—California. 3. Grasses—California.
   I. Harlow, Nora. II. Jakob, Kristin. III. Raiche, Roger.

SB413.L7W46 2003
635.9'34'09794--dc21

                                                     2003056417

13 12 11 10 09 08 07 06 05 04
10 9 8 7 6 5 4 3 2 1

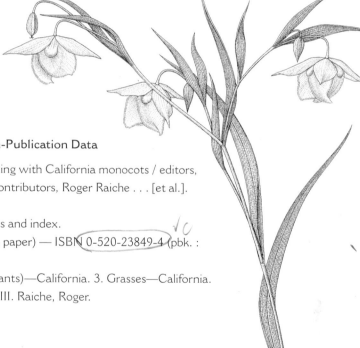

Cover photo: Iris douglasiana *and* Sidalcea malviflora *with nonnative* Festuca *'Siskiyou Blue' (until recently believed to be
a form of* F. idahoensis*) in the Berkeley, California,
garden of Jenny Fleming.* (SAXON HOLT)

DEDICATION

We dedicate this book to Lester Hawkins
and to Robert Ornduff. Lester was a
master plantsman, garden designer, and
co-owner of Western Hills Nursery in
Sonoma County. He inspired this work to
be begun in the early 1970s. Bob was a
green-thumb botanist and director of the
University of California Botanical Garden
for almost two decades. His support, both
personally and through the Stanley Smith
Horticultural Trust, helped to bring this
project to completion.

PLANTS IN THE WILD

Plants transplanted from the wild rarely survive in gardens. Harvesting seeds, bulbs, and other propagating material depletes the sources on which plants depend for reproduction and survival in their native habitats and contributes to plant rarity and extinction in the wild. It is also against the law to collect plants on public lands without a permit.

Although you may have to search for them, plants, seeds, and bulbs of species described in this volume are available from native plant nurseries, botanic gardens and arboreta, specialty seed suppliers, and native plant societies. Don't collect or disturb plants in the wild. Observe, photograph, and enjoy them where you find them.

CONTENTS

Iris douglasiana. (BART O'BRIEN)

PREFACE

This volume is the product of a native plant study group formed in the mid-1970s at the suggestion of nurseryman and landscape designer Lester Hawkins, who was constantly searching for ways to make botany and the science of horticulture more relevant to the interests and needs of gardeners. The study group met more or less regularly for at least five years in various locations, including the California Academy of Sciences, the Saratoga Horticultural Trust, Strybing Arboretum's Helen Crocker Russell Library, and eventually in group members' homes. At first these meetings were primarily an opportunity to exchange information and ideas and to meet people knowledgeable about native plants. Soon there was talk of compiling that information into a publication that could be shared with others. The California Native Plant Society (CNPS), which has published many important works on the unique and diverse California flora, expressed interest in developing and promoting such a book.

The initial focus of the study group was on cultivated varieties of native plants—selected or bred forms of wild plants grown primarily for their ornamental value. Anticipating a need for water-conserving plants and for plants adapted to the California environment, the group sought to assemble, and ultimately to disseminate, the knowledge and experience of group members in growing California native plants in gardens. The group discussed plants genus by genus and species by species, taking notes on their deliberations and observations. Then, working from these notes, group members began to write up each genus.

Beginning with the monocotyledons seemed a reasonable choice, since they are a clearly identifiable group and are not so numerous as to be daunting. Future volumes, it was assumed, would address flowering shrubs and vines, perennials and small shrubs, trees, and annual wildflowers. Publication of the monocots book eventually succumbed to other priorities, but it was not forgotten. In the mid-1990s, editor Nora Harlow was brought on board by the CNPS Publications Committee to help prepare the manuscript for publication. Authors worked to refine and update their sections. Again, however, other projects intervened, and it was not until the late 1990s, with the involvement of former CNPS publications chair Phyllis Faber and the University of California Press, and with financial support from the Stanley Smith Horticultural Foundation, that the book finally moved toward publication. Additional funding was received from the California Horticultural Society, the East Bay Chapter of CNPS, the North American Rock Garden Society, Gerd Jakob, and Robert Ornduff. We are grateful for the contributions of these organizations and individuals.

There are other books on the cultivation of native California plants, many more now than when this project began. What distinguishes this volume, and what fueled the group's efforts over the three decades of its preparation, is the vast and varied experience of its contributors. This book was produced by Jenny Fleming, Linda Haymaker, Kristin Jakob, Ron Lutsko, Jr., Elizabeth McClintock, Roger Raiche, Wayne Roderick, Suzanne Schettler, Jacob Sigg, Nevin Smith, and Caroline Spiller. The sections on alliums, brodiaeas, and fritillaries were written by Roger Raiche. Roger also authored the chapter on grasses and grasslike plants, with early contributions by Rosamond Day and the addition of a section on cattails by Jake Sigg. Jake prepared the chapter on agaves, yuccas, and nolinas as well. Nevin Smith prepared the section on calochortus. Ron Lutsko, Jr. was responsible for the descriptions of lilies, erythroniums, and trilliums. Suzanne Schettler wrote the section on irises and helped draft the glossary with Elizabeth McClintock. The section on lilylike plants was written by Linda Haymaker and Caroline Spiller. Kristin Jakob prepared all of the illustrations as well as the selected readings list, drafted the sections on resources, display gardens, and sources of plants, proofed and commented on the entire manuscript, and secured almost all of the funding for prepublication expenses.

Participating in early discussions were Dan Campbell, Barrie Coate, Rosamond Day, Betsy Flack, Roman Gankin, Lester Hawkins, Richard Hildreth, Barbara Ingle, Mary Meyer, Marshall Olbrich, Otto Quast, Marjorie Schmidt, Glen Schneider, Robert Smaus, Don Thomas, and Stewart Winchester. Others who contributed information in their specialties were Betsy Clebsch, Beecher Crampton, Charli Danielsen, Jim Dunne, Marge Edgren, Stan Farwig, Marge Hayakawa, Bob Hornback, Walter Knight, Barbara Menzies, Bart O'Brien, Warren Roberts, Jeff

Rosendale, Ed Smith, and Jimmy Vale. Peggy Fiedler reviewed early drafts and made recommendations that significantly altered the organization of the book. Carol Bornstein and Steve Edwards reviewed major portions of the manuscript and provided invaluable information and suggestions for revision that greatly improved the final product. Jenny Fleming, through her persistence, dedication, and unwavering belief in the value of the project, kept everyone moving forward toward completion over many years. To all of these individuals we owe our gratitude. Errors or omissions are, of course, the responsibility of the editors.

Nora Harlow and Kristin Jakob

California Counties

1 Alameda	21 Marin	41 San Mateo
2 Alpine	22 Mariposa	42 Santa Barbara
3 Amador	23 Mendocino	43 Santa Clara
4 Butte	24 Merced	44 Santa Cruz
5 Calaveras	25 Modoc	45 Shasta
6 Colusa	26 Mono	46 Sierra
7 Contra Costa	27 Monterey	47 Siskiyou
8 Del Norte	28 Napa	48 Solano
9 El Dorado	29 Nevada	49 Sonoma
10 Fresno	30 Orange	50 Stanislaus
11 Glenn	31 Placer	51 Sutter
12 Humboldt	32 Plumas	52 Tehama
13 Imperial	33 Riverside	53 Trinity
14 Inyo	34 Sacramento	54 Tulare
15 Kern	35 San Benito	55 Toulumne
16 Kings	36 San Bernardino	56 Ventura
17 Lake	37 San Diego	57 Yolo
18 Lassen	38 San Francisco	58 Yuba
19 Los Angeles	39 San Joaquin	
20 Madera	40 San Luis Obispo	

I. INTRODUCTION

Imagine, for a moment, a city or suburb in which residential front yards, commercial landscapes, parking lots, streetscapes, and public parks feature plants compatible with their natural surroundings, evocative of the local history, and synchronized with the seasons. Many of these plants may not be native to the geographic spot in which they grow, but all fit comfortably into a regionally appropriate landscape scheme. They look right, feel right, and connect us instinctively to the natural order of things. "A thing is right," wrote Aldo Leopold in *A Sand County Almanac* in 1949, "when it tends to preserve the integrity, stability, and beauty of the biotic community." We could add that a thing also seems right when it stirs collective memories of a place, building on the past and anticipating a congenial future. Native plants, skillfully integrated into a private garden or public space, have the matchless ability to do all of these things.

Why do we choose alien plants when there are so many native plants, better adapted to the climate and soils of the area and more likely to thrive without extra attention? The most commonly mentioned reason for the preference for nonnatives is that people tend to choose plants that remind them of home, and many Californians have migrated here from other parts of the world. Those other lands support many beautiful landscape plants, some of which require considerable effort and resources to keep them alive in California, while others have adapted so well to our climate and soils that they spill over into neighboring wildlands and crowd out native species.

Two other reasons for the prevalence of nonnative plants in California gardens have been the general lack of knowledge about California natives and the perception that natives are the province of experts, difficult to grow. Native plants also have been slow to appear in the nursery trade, so in the past many good landscape plants have not been widely available.

1

Over the past two hundred years, as more than a thousand nonnative plants were introduced into the California flora, about the same number of the state's native plants became rare or endangered and almost thirty became extinct.[1] Conversion of wildlands to agriculture and urban development have played a major role in the disappearance of the state's native plants, but anyone who clears a piece of land and creates a landscape entirely of nonnative plants contributes to the decline of native species.

There are several steps Californians can take to preserve our native plants. The most important may be to preserve entire habitats by setting aside wildlands where typical, rare, or endangered species occur in significant numbers. Botanic collections of representative species also help to ensure that these plants will not become extinct. And each of us can make a small but important contribution by including at least some native plants in our gardens, by encouraging nurseries to make attractive and suitable natives available in the trade, and by sharing our knowledge and appreciation of native plants with colleagues, friends, and neighbors.

This book was written to encourage gardeners, homeowners, landscape designers, and others to consider natives when designing, planting, and evaluating plans for landscapes in California. The authors, who write from personal experience in growing and propagating native plants, focus here on native monocots, a large and varied group of flowering plants that includes lilies, irises, grasses and grasslike plants, orchids, agaves, and even palms.

Before focusing on the plants themselves, it is useful to cover some basic information on monocots, plant rarity, and the sometimes confusing issues surrounding the classification of plants—the meaning of plant names and why they change.

What Is a Monocot?

With more than 230,000 described species grouped into about 400 plant families,[2] the world's flowering plants, or angiosperms, are the dominant group of plants on land and the major component of the vegetation in most natural ecosystems. Flowering plants occupy a greater range of natural environments than any other land plants, from tundra to tropics and from deserts to aquatic ecosystems. They exhibit an enormous range of life forms, from short-lived annuals to long-lived trees and shrubs.

Two major subdivisions of angiosperms have long been recognized: the dicotyledons or dicots (having two cotyledons or embryo "seed leaves"), which constitute about three-quarters of all flowering plants; and the monocotyledons or monocots (having a single cotyledon).[3] Besides the number of cotyledons, the two groups conventionally have been distinguished by the characters listed in the table below.

CHARACTERISTICS OF MONOCOTS AND DICOTS

MONOCOTYLEDONS	DICOTYLEDONS
embryo with 1 cotyledon	embryo with 2 cotyledons
flower parts usually in multiples of 3	flower parts usually in multiples of 4 or 5
leaves with parallel veins	leaves with reticulate ("net") venation
vascular bundles scattered (in stem cross section)	vascular bundles in a ring (in stem cross section)
no secondary growth by vascular cambium	secondary growth by vascular cambium
pollen monosulcate or monoculpate (1 furrow or pore)	pollen mostly tricolpate (3 furrows or pores)
adventitious root system from stem nodes or hypocotyl (no primary root)	persistent primary root from embryonic radicle

Although no single character can invariably distinguish monocots from dicots, these general distinctions are useful for horticultural and field identification purposes. In monocots, the numbers of petals, stamens, and other flower parts generally are divisible by three; the typically narrow leaf blades usually have a number of primary veins that run parallel the length of the leaf (the small connecting cross veins are inconspicuous); and the diffuse, fibrous roots arise adventitiously from nodes in the stem rather than from a primary ("tap") root as in most dicots.

Most monocots are herbs, and none forms true wood (stems do not thicken year by year through production of growth rings). Treelike monocots, such as palms, bamboos, and some yuccas, are supported by the thick-walled cells of a strong but nonwoody stem. Most monocots grow from bulbs, corms, or rhizomes; a few are vines. There are about 60,000 species of monocots in some sixty families, although any estimate of the number of families is speculative because classification of flowering plants, especially of the larger families, is in a state of flux.

Scientific Names of Plants

Scientific names consist of a genus, or generic name, and a specific epithet (often incorrectly called the "species name"). Together, the genus and the specific epithet make up the species name. Often the species name includes categories below the level of the species, such as subspecies (abbreviated "ssp.") and variety (abbreviated "var."). Plants also have a family name, which usually ends with the suffix -aceae, as in Poaceae, the grass family, or Liliaceae, the lily family.

The scientific names of plants are important because they identify the plant in question with some precision and clarity. Scientific names are far more reliable and more universally accepted than are the multiple "common" or vernacular names applied to plants, which differ from one region, supplier, or gardener to another. But scientific names do change, and at any given time there may be disagreement among botanists and taxonomists about the correct scientific name of a specific plant.

Plant Classifications

Currently, no general consensus exists regarding the taxonomy or classification of plants. Many California botanists today use the classifications set forth in *The Jepson Manual: Higher Plants of California*,[4] but it is important to recognize that there are other classification systems and that all classifications change over time. Taxonomy is a dynamic discipline, with classifications changing as new information on genetics, chemistry, and anatomical details becomes available and as new approaches to ordering that information become widely accepted.

Historically, plants have been classified, or grouped, as conceptually useful units (taxa) based primarily on observed similarities in certain physical characteristics. Flowering plants, for example, usually are grouped according to similarities in reproductive features (flowers and fruits) plus some vegetative and chemical characteristics.[5] Another system groups plants by common lineages—that is, by their presumed evolution from a common ancestor. Under this approach, a single lineage is separated into different taxa at points where two groupings are believed to have diverged as plants evolved under different circumstances.[6] As more information on plant characteristics becomes available (e.g., from molecular or chemical analyses), or as opinions about the relevance of particular features are modified, the designations of taxa may change.

For those more interested in knowing and growing plants than in the finer points of taxonomy, it is important only to understand that experts often have different ideas about how observed facts should be conceptually organized. In general, there tend to be taxonomic "lumpers" and "splitters"—those who propose greater inclusiveness in the categories they devise and those who propose categories that are more finely differentiated.

A prominent example of this is seen in the widely varying ways in which lilies and their relatives are treated in reference books. *The Jepson Manual* adopts a lumping approach, retaining a broad definition of the lily family (Liliaceae) as including lilylike plants as diverse as agaves and trilliums. The authors explain that they adopted this approach because "once any one of a large number of groups (such as the agave-like genera) is taken out [of a family], the justification for keeping other groups within the family is weakened."[7] Noting that some authors have adopted a less conservative (splitting) approach, they add: "If a modestly free hand in segregating lily-like families were allowed, in California we would have at least Agavaceae, Alliaceae, Aloeaceae, Asparagaceae, Calochortaceae, Convallariaceae, Liliaceae, Smilaceae, and Trilliaceae"[8]—nine separate families (or more) instead of just one.

In this volume we have attempted to take a neutral stance, consistently noting the family designations of *The Jepson Manual* but also, for many plants, mentioning other families to which they have been assigned by other authors. Because this book is first and foremost about growing plants, we group plants not by family but according to horticultural considerations. For example, although agaves and fritillaries are both traditionally placed within the Liliaceae, they generally are not combined in the garden or culturally managed in the same way, and so in this book they are discussed in separate chapters. Likewise, because garden uses of alliums and brodiaeas are more similar to each other than they are to those of yuccas and nolinas or even to those of erythroniums and trilliums, they are discussed separately, even though some taxonomic texts place them in the same family. Species names are consistent with those cited in the University of California, Berkeley, Jepson Flora Project's Online Interchange for California Floristics.

Rarity in Plants

Rarity and threat to survival vary among plant species, depending on the numbers of individual plants of the species, their natural range, and the pressures they face from the activities of humans (such as agriculture and urban development, off-road vehicles, or horticultural collection), animals (such as overgrazing), and competition from other plants (usually invasive exotics). There are many reasons why a plant may be considered rare, but in general a species is rare either because its natural habitat is limited (natural rarity) or because its habitat is being altered or has been largely destroyed.[9]

A major reason for plant rarity in California is that more than one-third of the state's native species are "endemics"—that is, they are restricted to a particular locality or habitat within the state. Most California endemics are rare or at least uncommon.[10] These plants are specialists—adapted to particular combinations of climate and soil—and many specialized plant habitats in the state are being, or already have been, substantially altered or destroyed. Most of California's annual and perennial grasslands and vernal pool habitats, for example, are gone.[11] Many more plants and plant habitats are seriously threatened and may disappear if current trends continue.

There are other reasons for plant rarity. Some specialist plants, such as those restricted to serpentine soils, are naturally rare primarily because their habitats are limited. Some plants are rare because their biology is such that few seeds germinate or seeds are poorly dispersed in the wild. These plants may never be other than rare unless, through deliberate propagation and distribution through the horticultural trade, they become widely grown by gardeners.

There is no statistically precise definition of plant rarity.[12] Federal and state laws and the California Native Plant Society (CNPS) recognize several levels of threat or endangerment. In general, "endangered" species are those considered to be in immediate jeopardy and likely to become extinct; "threatened" species are those likely to become endangered if not given special protection; and "rare" plants are those present in such small numbers throughout their range that they may become endangered if their environmental situation worsens.

Federal laws under which native plants are protected and managed include the federal Endangered Species Act, the National Environmental Policy Act, the National Forest Management Act, and the federal Clean Water Act. State laws include the California Environmental Quality Act, the California Forest Practices Act, the Natural Communities Conservation Planning Act, the Native Plant Protection Act, and the California Endangered Species Act. Federal law prohibits the destruction or "taking" of listed species on federal lands or on lands impacted by projects under federal control. State law prohibits the destruction of state-listed species under most circumstances.

CNPS maintains and publishes an inventory[13] of rare and endangered native plants that consists of five lists: List 1A includes plants presumed extinct in California; List 1B includes plants rare, threatened, or endangered in California and else-where; List 2 includes plants rare, threatened, or endangered in California but more common elsewhere; List 3 includes plants for which more information on rarity and endangerment is needed (a "review" list); and List 4 includes plants of limited distribution (a "watch" list). Plants on the watch list, though not considered rare, may be drastically declining in some areas, threatened with extirpation in some portion of their range, and/or show little ability to regenerate in many areas.

To further refine the designation of a plant's status, CNPS uses three independently scored factors: rarity (R: the number of individual plants and the extent of their distribution); endangerment (E: vulnerability to extinction); and distribution (D: the geographic range of the plant). In combination, the R-E-D code indicates the degree of rarity or endangerment, with 1-1-1 being the lowest degree of concern and 3-3-3 the highest.

The California native monocots with the most rare species are the alliums, lilies, fritillaries, erythroniums, and calochortus.[14] Many other genera include at least some rare or endangered plants. For example, twenty-seven native species of *Carex* currently are listed by CNPS as rare, endangered, or potentially endangered, as are eleven species of *Juncus* and more than thirty grasses. Most native orchids are rare, and few are available in the nursery trade.

About this Book

There are many more California native monocots than are discussed in this small volume. *The Jepson Manual* lists 1,156 species, subspecies, and varieties in 228 genera of native California monocots. Only about 250 of these are described and illustrated here.

The plants featured in this book are those bulbs, grasses, and other native monocots that can be grown successfully in gardens and are reasonably available in the nursery trade or from plant societies or seed exchanges. Threatened, endangered, or otherwise sensitive plants that are available as nursery-propagated plants, bulbs, or seeds are listed, and their conservation status is clearly identified. Gardeners should grow these plants only if they can be obtained from reputable suppliers. Very rare plants and plants not known to be available in the trade have been omitted from this book. Although some of these are California's most beautiful native monocots, until nursery-propagated plants and seeds become available, or until cultivars are selected or bred that do well in gardens, these plants should be appreciated only in the wild.

In the chapters that follow, three broad groups of plants are described: bulbs and bulbous plants, grasses and grasslike plants, and a group of mostly subtropical succulent or xerophytic monocots. Each chapter begins with some introductory information useful to gardeners, including general cultural similarities or differences within the group and tips on garden cultivation and propagation.

Individual plant descriptions provide greater detail on each species, including distribution or range and preferred habitat in the wild, cultural preferences and tolerances in the garden, and features that distinguish the plant from similar species.

Cultural preferences suggest the conditions under which the plants are most likely to thrive, although many will succeed in other circumstances with a little coddling. Preferred elevations and bloom times are relative. In general, "low elevation" means sea level to 1,000 feet, and "high elevation" means over 6,000 feet. The designation "early spring" means only that the plant will tend to bloom earlier than a plant that blooms in "late spring, early summer." It is impossible to specify months of blooming without experience in growing the plant in a particular location. A plant that blooms in one garden in April may not bloom until late May in a garden only five miles away at a higher elevation, in shade rather than sun, or in open ground rather than free-draining containers. The natural range and distribution of each plant also are given in some detail because this information suggests where and under what conditions a plant may be most successfully cultivated.

Cultural preferences are summarized in a table at the back of the book entitled "Site Preferences of California Native Monocots." Nurseries and seed suppliers that carry native monocots, selected publications, and public gardens where mature plants can be viewed are listed in the sections entitled "Resources," "Selected Readings," and "Display Gardens," respectively. Most terms are defined in the glossary, and plants can be located in the index by both scientific and common names.

If this small volume serves its intended purpose, California gardeners and garden designers will begin to incorporate more native bulbs, grasses, and other monocots in their landscapes. They will urge nurseries to carry these plants, and nursery owners will encourage propagators to make them more available. Propagators will experiment with more species and varieties, seeking to identify those best suited to garden cultivation. As native monocots become more widely available, and as their contributions to the cultivated landscape become better known and appreciated, future volumes may include dozens of additional garden-worthy species.

ENDNOTES

1 *Inventory of Rare and Endangered Vascular Plants of California* (Sacramento: California Native Plant Society, 2001) (hereafter referred to as *Inventory*); Michael Barbour et al., *California's Changing Landscapes: Diversity and Conservation of California Vegetation* (Sacramento: California Native Plant Society, 1993).

2 Wendy B. Zomlefer, *Guide to Flowering Plant Families* (Chapel Hill: University of North Carolina Press, 1994). Under two of the major classification systems, flowering plants are divided into 437 families (Thorne) or 387 families (Cronquist). See R. F. Thorne, "Classification and Geography of the Flowering Plants," *Botanical Review* 58 (1992): 225–348; and A. Cronquist, *The Evolution and Classification of Flowering Plants* (Bronx, N.Y.: New York Botanic Garden, 1988).

3 Dicotyledons and monocotyledons were first given formal taxonomic standing in 1682 by John Ray in *Methodus plantarum nova*, and were popularized by the French botanist Antoine Laurent de Jussieu in *Genera Plantarum* (1789), a work that gradually replaced the system of plant classification devised by Linnaeus.

4 James C. Hickman (ed.), *The Jepson Manual: Higher Plants of California* (Berkeley: University of California Press, 1993).

5 Zomlefer, *Guide to Flowering Plant Families*.

6 Ibid.; R. M. T. Dahlgren, H. T. Clifford, and P. F. Yeo, *The Families of the Monocotyledons* (Berlin: Springer-Verlag, 1985).

7 Hickman (ed.), *Jepson Manual*.

8 Ibid. The authors chose a conservative, "broad brush" approach to classification in part because the volume is intended to be useful to a wide audience, including amateur botanists and gardeners. Plant keys and descriptions deliberately emphasize features visible with little or no magnification. Taxonomic differences in chemistry, cytology, or physiology are either ignored or noted but not fully treated.

9 *Inventory;* Peggy Fiedler, *Rare Lilies of California* (Sacramento: California Native Plant Society, 1996); E. Rabinowitz, "Seven Forms of Rarity," in *Biological Aspects of Rare Plant Conservation*, ed. H. Synge (New York: John Wiley, 1981); Peggy Fiedler and J. J. Ahouse, "Hierarchies of Cause: Toward an Understanding of Rarity in Vascular Plant Species," in *Conservation Biology: The Theory and Practice of Nature Conservation, Preservation, and Management*, ed. P. L. Fiedler and S. K. Jain (New York: Chapman and Hall, 1992).

10 *Inventory.*

11 Ibid., p. 1. Annual and perennial grasslands in the Central Valley occupy only 1 percent of their former extent, and 90 to 95 percent of the state's vernal pools are gone.

12 Peggy Fiedler, "Rarity in Vascular Plants," pp. 2–3 in *Inventory.*

13 See above, note 1.

14 Fiedler, "Rarity in Vascular Plants."

Trillium chloropetalum (FRAN COX)

II. BULBS AND BULBLIKE PLANTS

The enormous variety of bulbs and bulblike plants is one of the most remarkable features of the native California flora. Although habitat loss has greatly reduced their numbers and restricted their range, there still are places where hundreds of thousands of fritillaries carpet the ground with their delicate pink, white, yellow, or purple flowers, where wet meadows burst forth with a sea of blue camas flowers, where masses of wild onions and brodiaeas decorate montane meadows or dry, rocky plains, or where the elegant and astonishing yellow-orange flowers of leopard lilies rise as tall as one's head.

Many bulbs and bulblike plants are well suited to California gardens. While most require a dry summer dormancy, some will tolerate a bit of summer water if soil drains well, and most will survive the droughts that can be counted on to occur periodically in California. Native bulbs selected for their suitability to the site can punctuate a grassy meadow or a summer-dry border with spectacular seasonal displays, while requiring little more than admiration from the gardener.

What Is a Bulb?

Many monocots, loosely referred to as bulbs, grow from swollen underground stems or modified leaves that serve as food storage organs and enable the plant to survive dormancy of various lengths of time. These fleshy storage structures include bulbs, corms, tubers, tuberous rootstocks, and rhizomes.

Bulbs. True bulbs are composed of a much-reduced stem at the base of modified leaves. The basal stem, a small disk- or cone-shaped structure called the basal plate, consists of hardened tissue that is visible when the bulb is cut lengthwise. The onion used in cooking is a good example of a true bulb. The modified

11

leaves, which are the edible part of the onion, arise from the upper surface of the basal stem. They are more or less fleshy and overlapping, and they have no chlorophyll but contain stored food. During the growing season, the basal stem of the onion bulb gives rise, from its upper surface, to the shoot. Green leaves and shoot are the aboveground parts of the onion plant. From its lower surface, the basal stem of the onion bulb produces adventitious roots that anchor the bulb in place.

True bulbs may develop miniature bulbs, known as bulblets, which when grown to full size are known as offsets. Offsets can be separated from the parent bulb and replanted. The number of growing seasons required for offsets to reach flowering size depends on the kind of bulb and the size of the offset.

There are two kinds of bulbs. Tunicate bulbs have broad, closely overlapping modified leaves. The outermost leaves are thin and dry and form a papery or membranous coat called the tunic, which envelops the fleshy inner leaves. California natives with tunicate bulbs include *Allium*, *Calochortus*, *Camassia*, *Stenanthium*, and *Zigadenus*.

Nontunicate (or imbricate) bulbs, which lack the papery membranous covering, have fleshy scales that press against each other but usually do not overlap. Nontunicate bulbs native to California include such genera as *Lilium* and *Fritillaria*. Fritillaries also sometimes produce small offsets called "rice grain" bulblets. Lateral buds or bulblets may arise among the fleshy scales of the underground bulb and are a means of vegetative reproduction. Aerial bulblets, called bulbils, sometimes arise on the aboveground stem in the axils of the uppermost leaves. The native leopard lily, *Lilium pardalinum*, occasionally produces bulbils in the axils of basal leaves.

Corms. Corms are often referred to as bulbs, although the two are easily differentiated. A corm is a solid, more or less rounded or somewhat flattened, fleshy underground stem in which food is stored. The modified leaves surrounding the corm are not fleshy as are those of true bulbs, and they do not contain stored food. Corms reproduce by developing new corms above or at the sides of the parent corm. California natives with corms include such genera as *Bloomeria*, *Brodiaea*, *Dichelostemma*, *Erythronium*, and *Triteleia*.

Corms produce new corms on top of old, withered corms. Miniature corms, called cormels or cormlets, are produced between the old and new corms. These can be separated from the parent corms and planted out or stored over the winter for planting in spring. New corms usually produce flowers the first season, while cormels require two to three years to reach flowering size.

Tubers. Some California monocots grow from tubers or tuberous rootstocks, the tips of which eventually form new plants. *Clintonia*, *Scoliopus*, *Trillium*, and *Veratrum* species grow from short, thick rootstocks.

Rhizomes. A rhizome is a more or less horizontal modified stem that contains stored food. Rhizomes may grow underground or on the soil surface. Adventitious roots grow from the underside of the rhizome, and leaves and flower stalks arise from the upper surface. California monocots with rhizomes include such genera as *Disporum*, *Iris*, *Lysichiton*, *Maianthemum*, *Sisyrinchium*, *Smilacina*, and *Streptopus*.

This chapter describes some California bulbs and bulb-like plants that may adapt to garden cultivation, including lilies and lily-like plants, calochortus, trilliums, alliums and brodiaeas, irises, and orchids. Yuccas and agaves (Agavaceae), which grow from rhizomes, and nolinas (Nolinaceae), which grow from a partially underground woody stem or caudex, are discussed in the chapter on succulent and xerophytic plants because of their distinctive horticultural uses and needs.

LILIES AND LILYLIKE PLANTS

The lily family, or Liliaceae, once included an extraordinarily diverse group of plants, with about 300 genera and more than 4,500 species worldwide in an astonishing array of forms. Some major reference books still adopt this comprehensive approach,[1] but most botanists today view the lily family much less broadly.[2] Although there continues to be disagreement about the classification of plants formerly known as lilies, there is a growing consensus that the Liliaceae, narrowly construed, comprises ten to twenty genera and fewer than 450 species.[3] Only three of these genera, *Erythronium*, *Fritillaria*, and *Lilium* (the "true" lilies), include plants native to California.

Some other plants traditionally included in the Liliaceae are described in the section on lilylike plants, along with mention of the families in which they now tend to be placed.

The Lily Family

The lily family includes some of California's most beautiful perennials, coveted by gardeners throughout the world. Many, if not most, California lilies are endangered, threatened, rare, or uncommon in the wild for a variety of reasons.[4] Some are naturally restricted to a few localized populations or specialized habitats, but many more are threatened by competition from exotic species or by human activities, including horticultural collection and habitat loss or degradation by timber harvesting, development, vehicular traffic, or overgrazing.

Plants in the lily family create some of California's most beautiful wildflower displays. Many people make annual spring pilgrimages to see fields of fritillaries, but fewer seek out the dry, scrubby woodland or chaparral environments where erythroniums are found. Given the diversity of their habitats, observing native lilies in the wild can be truly inspirational, and this is perhaps the most satisfying way to appreciate these unique and beautiful plants.

Plants in the lily family occur in a variety of natural ecosystems and can be found in many different circumstances with respect to soil, moisture, temperature, and exposure. They grow along shaded streambanks; in sunny, wet meadows; on dry, rocky slopes and ridges; in the dappled shade cast by dry chaparral or woodland plants; and on austere serpentine barrens devoid of almost any other plant life. Although one might conclude that there must be a species that fits every garden location, this is not the case. Many California lilies have exacting requirements, which may be difficult or impossible to duplicate in gardens.

When considering this group for use in the garden, it is best to concentrate on those proven to be tolerant of garden conditions. Rather than attempting to grow marginally tolerant species, we should seek out those that have been cultivated by gardeners with some success. The sections that follow feature those native lilies and lilylike plants that may succeed in gardens if their cultural requirements are met.

ERYTHRONIUMS

Erythroniums, or fawn lilies, are spring-flowering, deciduous, bulbous perennials that inhabit humus-rich soils of the forest floor, heavy, stony soils of chaparral or dry woodland, or damp, fine-textured soils at the edge of mountain meadows. *Erythronium* belongs to the Liliaceae, along with *Fritillaria* and *Lilium*.[5]

Erythroniums are often considered difficult to maintain in gardens, but some are quite easy in woodland settings, shady perennial beds, sunny rock gardens, or containers. Plants arise from elongated bulbs one to three inches long. The four- to eight-inch-long basal leaves, which begin to uncurl in late winter or early spring, typically occur as an opposing pair lying almost flat on the ground. They are glossy green, and in some species are mottled with shades of pale silvery green or violet-brown. Slender stems, which appear a few weeks after leaves emerge, rise from four to fifteen inches above the leaves and are topped with delicate, nodding flowers in white, yellow, pink, rose, or lavender. Each stem supports from one to seven flowers, and virtually all species are showy in bloom.

The western United States contains most of the twenty-five species in this delightful genus, the others occurring in the eastern United States, Europe, and Asia. California has fourteen native erythroniums, most of them in northern parts of the state, with a particularly heavy concentration in the North Coast Ranges and the Siskiyou Mountains.

Growing Erythroniums

Erythroniums vary in their adaptability to garden conditions. Most prefer at least a partially shaded site, humus-rich, well-drained soil, and a touch of summer moisture, although some thrive in full sun in gravelly soil with little or no organic material. Most do best where the ground freezes in winter. Many can be grown in pots if they are not allowed to become overheated. As with other lilies, the tiny bulbs may eventually pull themselves down to the bottom of the pot, but they also make offsets that will bloom.

A handful of species, notably *Erythronium californicum*, *E. citrinum*, *E. helenae*, *E. hendersonii*, *E. multiscapoideum*, *E. revolutum*, and *E. tuolumnense*, are relatively easy to cultivate in gardens, provided their requirements are met. Plants from high elevations have proven virtually impossible to please under garden conditions and are not included here.

Bulbs and seeds occasionally are available from specialty nurseries and seed suppliers. Bulbs should be planted in early fall to allow them to develop roots before cold weather sets in. Set bulbs three to six inches apart with the base of the bulb about four or five inches below the soil surface. Plants need some moisture while they are actively growing and blooming. They will go dormant in summer.

Erythroniums also can be grown from seed. Seed is best sown in early fall in damp, gritty soil or potting mix with some garden soil added. Seeds must be allowed a drying period of a few weeks before sowing to keep them from rotting; however, viability may be diminished by prolonged dry storage that extends into winter. Although seed sown in early fall will tend to germinate in spring, germination may be delayed a full year. Seedlings look like grass the first season after germination.

Since growth is slow, seedlings should be kept in the same pots for at least two years. When the broader leaves of maturing plants appear, usually in the third year, plants can be set out in the garden in a spot with good drainage. Blooms may appear in the third season but are more typical in the fourth or fifth year from seed. Plants should be disturbed only when they become crowded (no more than once every three to four years). They can be lifted and divided after leaves have died down.

Native erythroniums have the elegance required for close-up inspection, yet, when planted in drifts, they are showy enough to attract attention from a distance. These diminutive and charming members of the lily family blend harmoniously with other woodland plants such as anemones, dicentras, cyclamens, ferns, hound's tongues (*Cynoglossum* spp.), polemoniums, and inside-out flower (*Vancouveria hexandra*). For contrast and counterpoint to the visual qualities of the chaparral garden, use them among bunchgrasses or beneath upright manzanita, scrub oak, or redbud.

Erythroniums for the Garden

The following erythroniums are some of the most likely candidates for garden cultivation.

Erythronium californicum, California fawn lily
Prefers: low to mid-elevation, shade, some summer moisture, good drainage
Accepts: summer dryness
Blooms: spring
One of the more easily grown species of *Erythronium*, California fawn lily is six to twelve inches tall, with bronzy marbled leaves. Flowers range from paper white with pale centers to ivory white with deep orange or mahogany etchings about the throat. This plant is quite tolerant of summer dryness. It inhabits dry chaparral, foothill woodland, and coniferous forest in the Klamath and North Coast Ranges of California, and in some locations can be seen by the thousands on hillsides in chaparral.

Erythronium 'White Beauty' is an offsetting clone of *E. californicum* selected by Carl Purdy in the early 1900s. Plants are eight to ten inches tall and have large white flowers with a narrow, brownish yellow throat and leaves with decorative brown flecks. This is one of the best erythroniums to start with, as it is dependable and free-flowering.

Erythronium californicum

Erythronium citrinum, lemon-colored fawn lily
Prefers: low to mid-elevation, shade, dry, good drainage
Blooms: early spring
Native to dry woodlands and shrubby slopes, often on serpentine, in the Klamath
Ranges of northwestern California and southwestern Oregon, this lovely fawn lily
has narrow, wavy-margined leaves mottled with brown or white and white flowers
with a yellow base that turn pinkish in age. It is relatively easy to grow, and seeds
and bulbs are occasionally available from specialty seed suppliers and botanic garden plant sales.

Erythronium helenae, St. Helena fawn lily

Prefers: low to mid-elevation, part shade, dry, good drainage

Accepts: full sun near coast

Blooms: spring

St. Helena fawn lily ranges from four to twelve inches tall and bears pure white flowers with golden anthers and a generous golden yellow throat. Leaves are heavily marbled. This plant inhabits stony serpentine or volcanic soils in dry chaparral or open woodlands at low to mid-elevations in the region of Mt. St. Helena in Napa and Sonoma Counties. Tolerant of summer dryness and some heat, colonies of this plant have been seen in full sun on a harsh stony site where summer temperatures exceed 100 degrees F. In the garden, plants multiply readily by vegetative reproduction.

Because of its limited distribution, *Erythronium helenae* is on the CNPS "watch" list (List 4). Bulbs and seeds occasionally are available from specialty nurseries and seed suppliers.

Erythronium hendersonii,
      Henderson's fawn lily

Prefers: mid-elevation, part shade, dry,
      good drainage

Blooms: early spring

Uncommon in dry woodlands and openings in the Klamath Ranges of northwestern California and southwestern Oregon, Henderson's fawn lily bears lovely pale violet to pink flowers, darker toward the tips, with a deep purple zone at the base, sometimes surrounded by a touch of yellow. Leaves are attractively mottled. Readily adaptable to garden cultivation, this is one of the loveliest erythroniums. On the CNPS List 2, it is rare in California. Bulbs and seeds are available from specialty nurseries and seed suppliers.

Erythronium hendersonii. (ROGER RAICHE)

Erythronium multiscapoideum, Sierra fawn lily
Prefers: low to mid-elevation, sun or shade, dry, good drainage
Accepts: full sun near coast
Blooms: early spring
Sierra fawn lily is four to six inches tall, with dark, strongly marbled, veined leaves and good-sized creamy white flowers with pale yellow to golden centers. The flower stem splits below ground, appearing to bear only one flower on each stem. Easily grown in gardens, it prefers a rocky site with good drainage and a dry summer dormancy; it spreads slowly by underground stolons, eventually forming dense colonies.

This erythronium is found on shrubby or wooded north-facing slopes in the summer-dry foothills of the Sierra Nevada and the Cascade Ranges.

Erythronium revolutum, pink fawn lily, mahogany fawn lily
Prefers: low to mid-elevation, some summer moisture, shade, good drainage
Blooms: spring

Pink fawn lily has large flowers in exquisite combinations of rose-pink with yellow or orange etchings on the throat. It is found at the edges of streams, along bog margins, and in damp woods in the Coast Ranges of northwestern California north to British Columbia. In the garden, plants prefer soils high in organic matter and may appreciate some summer water. Rare and endangered in California, the species is on the CNPS List 2. Bulbs and seeds occasionally are available from specialty nurseries and seed suppliers.

Some plants have white or yellow flowers. One with particularly rich pink flowers, broadly spreading petals, and darkly mottled leaves is offered commercially as *Erythronium revolutum* **'Pink Beauty'** or

Erythronium revolutum. (ROBERT CASE)

sometimes as *E. revolutum* var. *johnsonii*. Flowers are carried six to fifteen inches above the deep green, marbled foliage.

A cross between *Erythronium revolutum* and *E. tuolumnense* has produced two fine garden cultivars: **'Kondo'** has twelve-inch stems and nodding golden yellow flowers with a distinctive brown ring. **'Pagoda'**, with nodding yellow flowers on ten- to twelve-inch stems and nicely speckled leaves, is one of the most commonly available erythroniums.

Erythronium tuolumnense, Tuolumne fawn lily
Prefers: mid-elevation, part to full shade, some summer moisture, good drainage
Blooms: early spring

Tuolumne fawn lily has large, solid green leaves with wavy margins and buttercup-yellow flowers. Dependable and easy to grow in some shade with occasional summer water, this plant is native to open woodlands in the foothills of the Sierra Nevada, especially in Tuolumne County, on shady banks in moist, humus-rich soils. Dormant in late summer, it reappears in early winter. Bulblets are readily produced. Rare in California and elsewhere, it is on the CNPS List 1B. Seeds and bulbs are available from specialty nurseries and seed suppliers.

FRITILLARIES

Fritillaries are charming bulbous plants in the lily family (Liliaceae), along with the genus *Lilium*, which contains the true lilies, and *Erythronium*. There are eighty to 100 species of *Fritillaria* in the northern hemisphere. About twenty species occur in California, and half of these are endemic to the state. Some are quite difficult to cultivate in gardens.

Fritillaries grow from fleshy scales that are clustered into a bulb. In some species the bulbs are surrounded by numerous "rice-grain" bulblets that can mature into bulbs. In late winter or early spring immature bulbs, detached scales, or rice-grain bulblets begin to produce juvenile foliage.

In some species the juvenile foliage consists of a solitary broad leaf resting on the soil surface. This leaf usually becomes larger each year until the bulb is mature. A mature bulb produces a flowering stalk, but never a juvenile leaf from the same bulb. Mature foliage is confined to the top half of the flowering stalk, typically occurring in whorls for part of the length and gradually diminishing upward. In other species juvenile foliage initially consists of a single leaf, and as the plant matures additional leaves form a basal rosette.

The flowering stalk arises from the center of the rosette, with additional leaves distributed along its length. Depending on the species, flower stalks vary in height from four inches to three or four feet, and usually appear from early to late spring, depending on elevation. Within a given species, the height of the flower stalk depends on cultural conditions as well as genetics.

California fritillaries bear some of the most beautiful flowers in the genus. With a few exceptions, flower color in other regions of the world tends toward greenish brown. In California we find plants with flowers of gleaming white, bright pink, rich scarlet, pure yellow, deep blackish purple, and various checkered, spotted, and patterned polychromes in green, chartreuse, umber, and plum. The flowers are pendulous, a floral habit admired in almost any genus.

Fritillaries are found throughout the state, except in desert areas. Occasionally they make a massive display over several acres, but more often they occur sporadically in dense colonies. They are found in several different habitats: open woodland and chaparral, open grassy hillsides and meadows, open talus slopes and rock scree, and dense oak woodlands.

Growing Fritillaries

Fritillaries grow in specialized habitats, and in the garden they require specialized culture. At their best they are not easy garden subjects, and some have defied all attempts at cultivation. Some fritillaries can be grown in the ground if their natural conditions are closely replicated, but most do best in well-drained pots, raised beds, or rock gardens. Woodland species such as *Fritillaria affinis* (*F. lanceolata*) may be grown in loose, humus-rich soil with filtered light and excellent drainage. *Fritillaria biflora* may be content in heavy clay that bakes in summer. *Fritillaria purdyi* may thrive in a fragmented, deeply fissured rock outcrop.

Because fritillary bulbs have no tunic, they dry out easily. In the wild the bulbs of some species are protected by the heavy clay soil in which they grow. In cultivation, care must be taken to ensure that the bulbs are not exposed to too much heat during their summer dormancy, especially when grown in pots.

Garden-collected seed and seedling-grown plants occasionally are available from specialty seed suppliers and nurseries. Botanic gardens and plant society sales can be sources for some seedling-grown plants. Unlike wild plants, which inevitably fail to adjust to the changes imposed by transplanting, plants grown from seed may succeed, provided their requirements for drainage and for dryness in dormancy can be met.

Fritillaries for the Garden

The following fritillaries are some of the best candidates for garden cultivation.

Fritillaria affinis, checker lily

Prefers: low to mid-elevation, shade or part shade, dry summer dormancy, excellent drainage

Blooms: spring

Checker lily is one of the most variable of California's fritillaries. Native to the Coast Ranges of California north through Washington and the San Juan Islands, it can be found in open grassland, scrub, or chaparral and on slopes in foothill woodlands. It grows from bulbs consisting of a few scales surrounded by numerous rice-grain bulblets.

Plants on serpentine soils or in other austere habitats tend to be twelve to eighteen inches tall and thin-stemmed. Petals tend to be small, often narrow, and the pendulous flowers are open and star-shaped. Flowers may be purplish brown or reddish to almost yellow, usually checkered with a contrasting color. Plants growing in richer soils in woodland or coastal grassland can be robust, sometimes to three feet tall with thick stems and large, pendulous flowers with squared "shoulders." Flowers typically vary from yellowish or greenish brown with much checkering to purplish black with little checkering. The more robust plants resemble true lilies (*Lilium* spp.) in foliage, with three to five whorls of green, lanceolate leaves on the upper half of the stem. Plants may produce up to a dozen flowers, but three to six is more common. They seldom set seed. Flowering varies from early winter to late spring, depending on genetics and environment.

Fritillaria affinis. (CHARLES KENNARD)

Fritillaria affinis

Bulb and rice-grain
bulblets

Checker lily is one of the easiest fritillaries to grow, preferring a light, well-drained soil with some humus. It is adapted to many soil types, will take sun in coastal regions, prefers light shade in the interior, and will naturalize if its cultural requirements are met. It is attractive on a lightly shaded slope with heucheras, goldback ferns (*Pentagramma triangularis*), milkmaids (*Cardamine californica*), and woodland stars (*Lithophragma* spp.).

Fritillaria biflora. (MARLIN HARMS)

Fritillaria biflora, chocolate lily, mission bells

Prefers: low to mid-elevation, coastal habitats, sun, dry summer dormancy, good drainage

Accepts: serpentine, clay soils

Blooms: late winter, early spring

Chocolate lily has bright green leaves irregularly crowded on the lower half of stout stems six to eighteen inches tall. Dark purple-brown, pendant flowers, often with whitish markings and a glossy finish, appear in late winter or early spring in mild climates. This plant is almost always found within several miles of the coast.

Native to open, grassy slopes and woodlands in both northern and southern coastal California, chocolate lily prefers good drainage in sun and is reasonably easy to grow in containers or in rock gardens. It is striking with white-flowered *Sisyrinchium bellum* or meadow foam (*Limnanthes douglasii*). Propagation by seed is easy.

Fritillaria eastwoodiae, Butte County fritillary

Prefers: low to mid-elevation, light shade, dry summer dormancy, good drainage

Accepts: serpentine

Blooms: early spring, spring

Butte County fritillary is an attractive and variable species native to foothill woodland and chaparral in the western Sierra Nevada. From a cluster of thick scales surrounded by many rice-grain bulblets, it produces a sturdy stalk twelve to twenty inches tall with narrow leaves in whorls on the upper stem.

Flowers are small, bell-shaped or more open and starlike, and usually nodding. There may be as few as three or as many as fifteen flowers. Petals vary from yellowish green, yellowish pink, or yellowish orange to orange-green, orange, or creamy scarlet.

Butte County fritillary is not easy to maintain in gardens, but it may succeed in a loose, humus-rich soil in light shade. It is occasionally available from seed exchanges and specialty seed suppliers.

Fritillaria lanceolata, see *Fritillaria affinis*

Fritillaria purdyi, Purdy's fritillary
Prefers: mid-elevation, dry summer dormancy, sun, good drainage
Blooms: spring
Uncommon on dry ridges, generally on serpentine, in northern California and Oregon, Purdy's fritillary is reasonably easy to grow in well-drained soils in sun or part sun. Its spring flowers are white or greenish white and spotted or streaked with purple or purplish brown and pink shading. It is on the CNPS "watch" list (List 4). Seeds sometimes are available from seed exchanges and specialty seed suppliers.

TRUE LILIES: THE GENUS *LILIUM*

Throughout history, gardens worldwide have been adorned with lilies. The elegant spires topped with waxy, sometimes fragrant flowers have appeared in perennial borders and as stately subjects of woodland and streamside gardens. Unfortunately, most of the twenty or so species native to California are difficult to grow, and a few have proved impossible in cultivation. Several, however, can be grown in gardens if their cultural requirements are met.

California lilies are striking plants architecturally. The single stem, two to six feet tall or more with tiers of whorled leaves, is topped by showy, often ascending flowers in colors ranging from white to pink, yellow, orange, or red. Habitats in the wild also are varied, from coastal marshes, streams, and damp woods to interior chaparral, dry woods, and mountain meadows. Each species is quite specialized in habitat and cultural requirements, and most are of limited distribution in the wild. They tend to fall generally into two broad cultural groups—wet growers and dry growers—but there is considerable variation within the groups, and in some cases the two groups seem to overlap.

California lilies add an enchanting, almost animate presence scattered throughout an open grove of trees or thrusting their colorful, waxy-petaled flowers out of masses of chaparral shrubs. In their most classic garden uses, drifts of lilies combine well with rhododendrons, ferns, aquilegias, inside-out flower (*Vancouveria hexandra*), and hound's tongue (*Cynoglossum* spp.), making an unforgettable display in the cool, dappled shade of the woodland garden.

More than half of the state's native lilies are considered rare, threatened, endangered, or sufficiently uncommon to be of concern. Those listed here are available from several sources in the specialty nursery trade or from botanic garden and plant society sales.

Growing Lilies

Lilies grow from scaly rhizomes, which can be purchased from specialty bulb suppliers. Plants also can be grown from seed. Seeds should be started in raised beds or pots. Excellent drainage is of utmost importance in growing most native lilies, and this is best achieved with combinations of sand, gravel, and some organic matter as the growing medium; with the possible exception of some forms of *Lilium pardalinum*, no clay, silt, or loam should be used in soil mixes.

Watering schedules are difficult to specify. In the wild, dry growers are found away from apparent water sources and, aside from fog in some regions, seem to receive no summer water. However, when this regime is duplicated in the garden, losses usually occur. Wet growers inhabit areas that are perennially damp, with moisture fluctuating during the year, usually peaking in spring and ebbing in fall. If consistent watering is practiced in the garden, plants usually rot in summer. Occasional but infrequent watering in summer may provide success with both wet and dry growers.

If correct soil and watering conditions can be provided, plants usually can be brought to flower from seed in one and a half to five years, depending on the species. Wet growers and dry growers with some potential for garden cultivation are described below.

Lilies for the Garden

The following species of the genus *Lilium* are good candidates for garden cultivation.

Lilium humboldtii, Humboldt's lily
Prefers: mid-elevation, part shade, dry summer dormancy, excellent drainage

Blooms: mid- to late summer

This stately lily bears nodding, maroon-spotted, golden orange flowers with dark red blotches and pendant stamens with orange to brown anthers. The flowers are on stout stems five to eight feet tall. The whorls of bright green leaves have wavy margins. Humboldt's lily is easy to grow, and, planted in large numbers, it can serve as a spectacular framework for the perennial garden. It is best planted in a bed of deep, well-drained soil, and it prefers part shade and no water after blooming.

Lilium humboldtii **ssp.** *humboldtii* has orange flowers with magenta spots; it is an uncommon species of yellow pine forests of the high Sierra Nevada and Cascade Ranges. *Lilium humboldtii* **ssp.** *ocellatum* has large, showy yellow-orange or golden orange flowers with prominent red spots, larger toward the tips. It is found along streams in coniferous forest and coastal chaparral or on dryish slopes in dense coniferous forest, mostly in southwestern California. Because of their limited distribution in the wild, both subspecies are on the CNPS "watch" list (List 4). Both are occasionally available from native plant nurseries or botanic garden and plant society sales.

Lilium kelleyanum, Kelley's lily

Prefers: mid- to high elevation, part shade, wet places, excellent drainage
Blooms: early summer

An uncommon lily from hillside seeps and streambanks in subalpine coniferous forests of the Sierra Nevada, Kelley's lily bears fragrant, rich golden yellow flowers on four- to six-foot stems in its native habitats, but grows about half as tall in cultivation. Difficult to maintain in gardens, especially in open ground, this mostly high-elevation lily requires excellent drainage and some moisture in dappled shade or morning sun. It may do best in raised beds or containers.

Lilium pardalinum, leopard lily

Prefers: low to mid-elevation, shade, some summer moisture, good drainage

Lilium pardalinum. (SAXON HOLT)

Lilium pardalinum

Accepts: full sun with water near coast
Blooms: early summer

Leopard lily is well suited to garden cultivation. An elegant plant to six feet tall, it bears several to a dozen or more large, pendulous, yellow-orange flowers with reddish brown to maroon spots and conspicuous anthers atop a sturdy stem. Native to streambanks and wet places in many parts of California, especially coniferous and mixed evergreen forests, it is easy to grow if given moisture and part shade. Near the coast, it may flower best with water in full sun.

Lilium pardalinum **ssp.** *pitkinense*, Pitkin Marsh lily, is native to freshwater marshes and wet meadows in Sonoma County. Three to six feet tall with yellow-green leaves, it bears large, showy flowers, red at the outer edge and yellow at the center with deep maroon spots. Rare and endangered in California and elsewhere, it is on the CNPS List 1B, but is reliably available from nursery-propagated stock.

Lilium pardalinum **ssp.** *wigginsii*, Wiggins's lily, is found along streams and in bogs in the Klamath Ranges, the western Siskiyous, and southwestern Oregon. It bears clear, golden yellow-orange flowers and persists well in gardens if given a cool location with excellent drainage and damp but not wet conditions after flowering. Wiggins's lily is on the CNPS "watch" list (List 4). Sometimes listed as *L. wigginsii*, it is occasionally available from specialty suppliers.

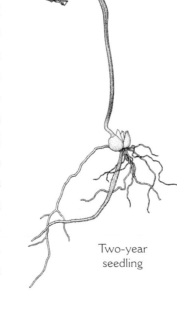

Two-year
seedling

Lilium pardalinum

Seeds

scales with bulblets

Seed pod

Lilium parvum

Lilium parvum, alpine lily
Prefers: mid- to high elevation, sun or part shade, wet places, good drainage
Blooms: summer
Native to wet meadows, willow thickets, and streams in coniferous forests in the northern and central high Sierra Nevada, alpine lily can bear dozens of outward- and upward-facing trumpet-shaped flowers in shades of orange with the throat beautifully marked with purple or maroon spots. The foliage rises in whorls up the sturdy three- to five-foot stems. Though not easy to grow, this lily may settle down and become established if planted in moist, humusy soil in dappled sun in a woodland setting.

From mid-elevations on the western slopes of the Sierra and more easily grown than the species, **Lilium parvum var. luteum** bears rich yellow flowers.

Lilylike Plants

The following plants traditionally have been considered members of the lily family, but more recently have been placed in other plant families by some taxonomists, as noted below. Major reference books such as *The Jepson Manual* and *Sunset Western Garden Book* still include them in the Liliaceae.[6]

Camassia leichtlinii, see *Camassia quamash* ssp. *quamash*

Camassia quamash, camas
Prefers: low to high elevation, moist soils, sun
Blooms: late spring, summer
Four species of *Camassia*, or camas, are distributed across North America. Only one species and two subspecies are native to California. Camas grows from bulbs, either singly or in clusters, in moist to wet, humusy soils, commonly in wet meadows, seeps, and damp forests and on stream banks from Marin County to

Camassia quamash. (WILLIAM FOLLETTE)

the northern Cascades and northeastern Sierra Nevada, southwestern Canada, Wyoming, and Utah. Some authors have placed *Camassia* in the Hyacinthaceae, along with *Chlorogalum* and *Hastingsia*.[7] Others have suggested that both *Camassia* and *Chlorogalum* belong in the agave family (Agavaceae).[8]

Camas is notable for its massive displays of beautiful blue flowers, which can cover entire meadows in late spring to mid-summer, giving the effect of a large lake or pond. Camas bulbs were eaten by Native Americans, who prepared them in various ways as a sweet and nutritious staple food.

The camas bulb is smooth and onionlike, growing to about an inch in diameter. The bulb produces linear, green strap leaves to two feet long and a flowering stalk from one foot to two feet high. The flowers are slender, six-petaled, and star-shaped on a slender raceme. Color ranges from white to blue or purple. After flowers fade, numerous black seeds form in dry, three-lobed capsules on the stalk.

Camas is effective combined with various meadow grasses, native columbine (*Aquilegia formosa*), leopard lily (*Lilium pardalinum*), sneezeweed (*Helenium bigelovii*), *Triteleia hyacinthina* and *T. peduncularis*, and buttercups (*Ranunculus* spp.). A meadow effect can be achieved by planting camas with medium-sized bunch-grasses. Camas is easily grown from seeds or bulbs. Bulbs are best planted where they can remain undisturbed for many years. Plants need summer moisture until the seedpods are fully mature.

The forms and colors of camas flowers are quite variable.

Camassia quamash ssp. breviflora grows in sunny, moist meadows at mid- to high elevations in the Sierra Nevada to Washington. Leaves are stout, wide, and glaucous, and the densely borne flowers with bright yellow anthers are one-half inch to three-quarters of an inch in diameter. Seeds occasionally are available from rock garden society seed exchanges.

Camassia quamash ssp. quamash grows in more varied locations than *C. quamash* ssp. *breviflora*, at both lower and higher elevations in northwestern California and the San Francisco Bay Area to southwestern Canada, Wyoming, and Utah. The flowers are larger, from three-quarters of an inch to an inch and a half in diameter, and the leaves are narrower. Plants of this species sometimes are offered as *C. leichtlinii*, a species not native to California.

Camassia quamash

Chlorogalum pomeridianum

Chlorogalum, soap plant, amole
Prefers: low to mid-elevation, sun or part shade, dry summer dormancy
Accepts: most soils, many habitats, heavy clay if dry in summer
Blooms: late spring, summer

Soap plants are stemless herbs with tufted basal leaves arising from a scaly bulb that produces a tall, leafless inflorescence. Native Americans and early settlers commonly used soap plants not only for soap but also for food, medicine, glue, brushes, and other purposes.

There are five species of *Chlorogalum* and several varieties native to California; some are rare. *Chlorogalum* is closely related to *Camassia*, and some authors have placed this genus in the Hyacinthaceae, along with *Camassia* and *Hastingsia*.[9] Others have suggested that *Chlorogalum* and *Camassia* should be placed in the Agavaceae.[10]

Chlorogalum pomeridianum is the largest of the soap plants native to California and the most often cultivated. Common where undisturbed in its native habitats, it is found from southern Oregon to San Diego County in mesic grasslands, on windswept ridges, and in more sheltered, shady sites in coastal sage scrub, chaparral, and foothill woodland.

The bluish green leaves of *Chlorogalum pomeridianum* form a wide-spreading basal rosette and are one foot or more long, with a ridged central vein and gracefully undulate margins. The large bulbs are covered with long, brown fibers. Although the individual flowers are small, masses of these plants in bloom are memorable, giving the effect of softly falling snow. The pearly white to pale lavender, six-petaled flowers are borne in loose clusters on tall, branching stems from late spring to mid-summer. Flowers tend to open at dusk; stalks cut during the day and brought indoors may bloom by evening.

One of the earliest bulbs to leaf out in late winter, *Chlorogalum pomeridianum* makes a quick cover in shade or sun, and it tolerates poor soils, including serpentine. If large numbers of the plant are not wanted, old flower stalks should be cut off after blooming, as masses of seed will sprout and grow. Overcrowding tends to inhibit blooming.

While *Chlorogalum pomeridianum* has a much wider range, **C. parviflorum**, small-flowered soaproot, is locally more common at lower elevations on dry, barren ridges or on bluffs in coastal sage scrub or grassland from Los Angeles County to Riverside and San Diego Counties and northern Baja California. Unlike those of *C. pomeridianum*, the white flowers of *C. parviflorum* open during the day. The bulbs are about half the size of those of *C. pomeridianum* and have a brown membranous rather than hairy coat. Leaves are about half or less the length and width of those of *C. pomeridianum*, and the flower stalks are about half as tall.

Clintonia, clintonia

Prefers: low to mid-elevation, coastal habitats, shade, moisture

Blooms: late spring, early summer

Six species of *Clintonia* are found worldwide, two in California. Both California species are uncommonly beautiful. Clintonias are propagated by seed or by division of rhizomes in early spring. Good companions are ferns, oxalis, trilliums, disporums, tiarella, and inside-out flower (*Vancouveria hexandra*). Some authors have placed *Clintonia* in the Uvulariaceae, along with *Disporum*, *Scoliopus*, and *Streptopus*.[11]

Clintonia andrewsiana. (STEPHEN INGRAM)

Clintonia andrewsiana, red clintonia, has elliptical, satiny, bright green leaves almost a foot long and up to five inches wide. Leaves sheath a short basal stem, gracefully curving upward and outward. In late spring or early summer, one or two stems rising above the leaves bear clusters of pink to deep rose flowers arranged in an umbel, followed by cobalt-blue berries that are every bit as arresting as the flowers. Red clintonia is found in humus-rich soils in shade at low to mid-elevations in coastal redwood forests in the Central and North Coast Ranges and the San Francisco Bay Area to southwestern Oregon.

Clintonia uniflora, bride's bonnet, has smaller leaves than red clintonia and usually bears a single, upright, fairly large, white flower in mid-summer. It is native from Tulare County to Lassen, Siskiyou, and Humboldt Counties and the Pacific Northwest eastward to the Sierra Nevada, generally in humus-rich soils of mid-elevation coniferous forests. A choice addition to the woodland shade garden in more northerly parts of the state, it has not proved successful as far south as the San Francisco Bay Area.

Disporum, fairy bells

Prefers: low to mid-elevation, shade, moisture, good drainage

Accepts: dry shade with northern exposure near coast

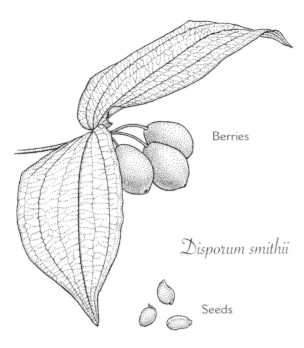

Berries

Disporum smithii

Seeds

Blooms: late winter, early spring
Fairy bells are small, leafy, herbaceous perennials with flowers that dangle in loose clusters. Crisp, ovate leaves, one inch to five inches long, have well-defined veins that parallel the wavy leaf margins. Plants emerge from slender, creeping rhizomes in late winter or early spring, when sharply pointed, bronzy red shoots appear out of bare ground. Within days the plant has leafed out, a bright green silhouette soon topped by an umbel-like inflorescence of delicate, bell-shaped, white or greenish flowers that are often hidden by the leaves; they are followed by brilliant orange-red berries. Some authors have placed *Disporum* in the Uvulariaceae, along with *Clintonia*, *Scoliopus*, and *Streptopus*.[12]

Two species of fairy bells, *Disporum hookeri* and *D. smithii*, are found in California. *Disporum hookeri* is restricted to northern California and the Sierra Nevada to Montana and western Canada, while *D. smithii* grows along the coast, especially in redwood forests, from the San Francisco Bay Area to British Columbia.

Propagation of disporums is easiest by division in early spring. Propagation by seed is slower. Fresh seed can be grown to first leaf in plastic bags on damp towels. Plants are available from nurseries specializing in native plants for woodland gardens. Columbines and heucheras are

Disporum smithii. (SAXON HOLT)

good companions for disporums. Even with supplemental water, plants will go dormant in late summer, and foliage can then be cut back.

Disporum hookeri, Hooker's fairy bells, is smaller than *D. smithii* and accepting of some dryness. In the San Francisco Bay Area it has been successfully grown in dry shade with north-facing exposure. Stems are up to two feet tall, and leaves are one to three inches long. Greenish white flowers are shaped like a top, with petal tips spreading open to reveal the stamens. The dainty flowers are best displayed in raised beds or in other situations where they can be viewed from below. Native to redwood and mixed evergreen forests, Hooker's fairy bells is also found on shaded dry slopes.

Disporum smithii, coast fairy bells, is larger and showier than *D. hookeri*, with stems to three feet tall, flowers nearly an inch long, and large, inflated, orange-red berries that resemble jalapeños. The creamy white bells are clipped off almost squarely at the base, and the petals spread only slightly. This plant is native to redwood and mixed evergreen forests.

Hastingsia alba, rush lily

Prefers: mid- to high elevation, sun, wet places
Accepts: some dryness in part shade
Blooms: summer

Rush lily bears dense racemes of small white flowers tipped with green, lilac, or pink. The three- to six-foot inflorescence is branched, giving an open, airy effect when the twenty to seventy white, yellow, or greenish flowers are in bloom in summer. The flat, elongated leaves rise one foot to almost two feet from a tunicate bulb.

This plant is native to wet meadows, bogs, rocky seeps, and redwood or mixed evergreen forests in the Cascade Ranges and the northern Sierra Nevada to southwestern Oregon. In the garden, rush lily can tolerate some dryness in part shade. It is easy to grow, and seeds germinate so readily that in wet places it can become invasive.

Hastingsia, with four species in California and Oregon, was segregated from *Schoenolirion*, a genus characteristic of the southeastern United States.[13] Some authors have placed *Hastingia* in the Hyacinthiaceae, along with *Camassia* and *Chlorogalum*.[14]

Maianthemum dilatatum, false lily-of-the-valley, two-leaf solomon's-seal

Prefers: low to mid-elevation, coastal habitats, shade, moisture
Blooms: spring

Maianthemum dilatatum

Maianthemum dilatatum. (ROBERT CASE)

Maianthemum is a north-temperate genus with three species, only one of which, *M. dilatatum*, is native to California. This low-growing perennial herb has heart-shaped leaves that form luxuriant, bright green carpets in damp woodlands. Shoots from a creeping underground rootstock appear in spring, accompanied by a pair of exquisitely veined and pleated leaves. At the top of the erect stem is a raceme of delicate white flowers. The bright green foliage is effective in the spring garden, and is accented by red berries after flowering. As summer heat moves in, plants go dormant until the next year. Generous water and shade will extend the display a bit. Snails and slugs love this plant, and it can become invasive.

False lily-of-the-valley is found in moist, shaded sites in coniferous forests near the coast from the San Francisco Bay Area to Alaska and Idaho. On San Bruno Mountain it grows among low-growing manzanitas and bunchgrasses with no tree cover, and in Humboldt County it is found in open brushland on sea bluffs. In the moist woodland garden it is especially attractive with ferns. The neat foliage makes it a good container plant. Propagation is by seed or by division in early spring before new growth starts.

Some authors have placed *Maianthemum* in the Convallariaceae, along with *Smilacina*.[15] Others place both *Maianthemum* and *Smilacina* in the Asparagaceae, along with *Asparagus* and *Convallaria*.[16]

Odontostomum hartwegii, Hartweg's odontostomum

Prefers: low to mid-elevation, sun, dry, clay soils
Accepts: serpentine soils, heavy clay
Blooms: spring

The genus *Odontostomum* has only one species, *O. hartwegii*, which grows five to fifteen inches tall, its flexuous stem sheathed in mostly basal linear leaves, and

Odontostomum hartwegii

with branches that end in wands of quarter-inch, toothed, cream-colored flowers clustered in a raceme. Flowering begins in spring as soils dry out.

A rather uncommon plant, Hartweg's odontostomum is native to and sometimes locally abundant in sun-baked clay soils of the inner North Coast Ranges and the Sierra Nevada foothills. This is a good plant for difficult garden situations— heavy clay and hot, dry sites with no supplemental irrigation. It can be combined with calochortus, brodiaeas, poppies, and other dry growers that thrive in adobe soils. It grows from corms.

Scoliopus bigelovii. (ROBERT C. WEST)

Scoliopus bigelovii, adder's-tongue, slink-pod
Prefers: low to mid-elevation, coastal habitats, shade, moisture
Blooms: late winter, early spring
This long-lived perennial herb is a memorable plant for woodlands, moist slopes, and shady rock gardens along the coast. Visible just three or four months of the year, it is grown more for its showy foliage than for its flowers, and it is an excellent plant for the winter garden. It should be placed so the unusual flowers can be seen. As the tightly wrapped leaves begin to unfold, green, tan, purple, or blackish flowers with narrow, erect petals appear on six-inch stalks above the leaves.

The odor of the flowers, which some people find unpleasant, is not noticeable except when plants are massed in large drifts. Flowers last a week or two, until the sepals drop; then the stalks with ripening seed pods elongate and twist toward the soil, where ants collect and disperse the seed. At the same time, the two shiny green leaves expand to display a peculiar purple-brown mottling, like spots on a fawn. A foot or more long and four inches wide in shade, the leaves have a pleated effect, which is pleasing for the remainder of spring until the leaves die down in summer.

Adder's-tongue is restricted to the coastal fog belt.[17] It is found within the range of coastal redwood forests in California from the Santa Cruz Mountains in Monterey County north to the Oregon border. This plant inhabits moist, shaded slopes, mostly in old-growth redwood forests and adjacent chaparral, but also in

Scoliopus bigelovii

coastal Douglas-fir or mixed evergreen forests. It is often found in ephemeral springs or seepages, but has been documented from drier, more open habitats such as rocky outcroppings on coastal headlands. Common understory companion plants include western sword fern (*Polystichum munitum*), red clintonia (*Clintonia andrewsiana*), western trillium (*Trillium ovatum*), redwood sorrel (*Oxalis oregana*), and evergreen violet (*Viola sempervirens*).

A rootstock four to five inches deep sends yellowish tan, fuzzy roots down to permanent moisture in humus-rich clays. Adder's-tongue is propagated by seed collected at the moment of opening and directly sown. Seeds must be kept moist until fall germination. Plants grown from seed require four years to bloom.

Some authors have placed *Scoliopus* in the Uvulariaceae, along with *Clintonia*, *Disporum*, and *Streptopus*.[18] Others believe this genus should be placed in the Trilliaceae.[19]

Smilacina, false-solomon's-seal

Prefers: low to high elevation, shade, moisture, good drainage
Accepts: summer dryness in shade along coast
Blooms: spring

Two species of *Smilacina*, or false-solomon's-seal, are found in California, *S. racemosa* and *S. stellata*. Plants grow from rhizomes. In contrast to "true" solomon's-seal (*Polygonatum* spp.), which has flowers that grow along the stem, flowers of false-solomon's-seal are borne in clusters at the end of the stalk. Blooming from spring to early summer, they are attractive with ferns, redwood sorrel (*Oxalis oregana*), fritillaries, lilies, and trilliums. Some authors have placed *Smilacina* in the Convallariaceae, along with *Maianthemum*.[20] Others place both *Maianthemum* and *Smilacina* in the Asparagaceae, along with *Asparagus* and *Convallaria*.[21]

Smilacina racemosa, large false-solomon's-seal, is an attractive two- to three-foot-tall woodland plant with rich, light green foliage, especially striking in deep shade. The inflorescences display delicate, frothy, white flowers in spring, producing a soft

and graceful effect somewhat like that of an astilbe. Clusters of red berries appear in late summer or early fall.

This species develops robust clumps two to three feet tall. Two or three plants grown together make a lovely accent planting. Some summer moisture may be needed in the garden, but in the San

Smilacina racemosa. (ROBERT CASE)

Francisco Bay Area, on north-facing slopes, the plants are found under native oaks and bays where soils dry out in summer. *Smilacina racemosa* is native to mixed evergreen and coniferous forests in the Cascade Ranges and the Sierra Nevada to Alaska and eastern North America.

Smilacina stellata

Rootstock

Smilacina stellata, slim false-solomon's-seal, is a good spring and summer groundcover for coastal woodland gardens. It bears white flowers in late spring through mid-summer, but is grown primarily for its attractive foliage. About one foot to two feet tall, it spreads by branching rhizomes in dry or moist soils in shade or, near the coast, in sun with north-facing exposure. Rhizomes are easily divided, usually at the time of fall rains.

Smilacina stellata is found from sea level to high mountains, but low-elevation plants are best for gardens. It occurs on moist slopes in mixed evergreen forests, along streams, in chaparral, and in coastal redwood forests. Plants grow quickly with winter rains, and young shoots rapidly form an abundant cover. It is a good seasonal groundcover under oaks in coastal areas because, if allowed to go dormant, it needs little summer water. In a summer-irrigated garden it can be invasive.

Smilax californica, California greenbrier
Prefers: low to mid-elevation, sun or shade, moisture
Blooms: early summer
This deciduous riparian perennial arises from a tuberlike caudex with a climbing or trailing stem densely clothed in short, needlelike prickles. It is found in full sun to shade on streambanks in mixed coniferous forests, often sprawling among brush or climbing high into trees, in the foothills of the northern Sierra Nevada and the Klamath and north Cascade Ranges to southwestern Oregon; it also grows along streams on alluvial flats in the Sacramento Valley.

Smilax californica

The attractive apple-green, heart-shaped to rounded, deeply veined leaves, two to four inches long, appear in spring and drop in late fall. The small white or greenish white flowers are inconspicuous. Not easy to propagate from seed and difficult to transplant, plants are not often available commercially. Some authors have placed *Smilax* in the Smilacaceae.[22]

Stenanthium occidentale, western stenanthium, western featherbells
Prefers: mid- to high elevation, shade, wet places, good drainage
Blooms: summer
A dainty and graceful plant to about a foot tall, western stenanthium has grasslike basal leaves and an inflorescence of narrow, bell-shaped, purplish or sometimes yellowish green flowers. It is native to moist streambanks and seeps at mid- to high elevations in the Klamath Ranges to western Canada and Montana. In the garden it should be placed where the delicate flowers can be seen up close and the plant can be kept moist at all times.

Western stenanthium is difficult to maintain in the open garden, and is perhaps best grown in containers. It not widely available, but seeds or bulbs occasionally can be obtained from specialty seed suppliers and seed exchanges.

Streptopus amplexifolius var. *americanus*, twisted stalk, American streptopus
Prefers: low to mid-elevation, shade, wet places, good drainage
Blooms: spring, early summer

Streptopus amplexifolius var. *americanus*

Twisted stalk is similar to disporum but with more erect stems, and the two are often placed in the same family (Uvulariaceae).[23] This perennial herb emerges from a long, horizontal rhizome. It has tiny, whitish, bell-shaped flowers on twisted stalks that grow from the leaf axils. Leaves clasp the stem alternately, zigzagging up the stem, giving the appearance of a pigtail. Flowers are followed by greenish yellow to red fruits.

Found along streams in moist coniferous forests, both coastal and montane, twisted stalk is not easy to grow in gardens, but it has succeeded in containers in shade. Propagation is from seed or by division in early spring. Plants and seeds are sometimes available from specialty nurseries and seed suppliers.

Veratrum, corn lily, false hellebore
Prefers: mid- to high elevation, shade, wet places, good drainage
Blooms: mid-summer
Veratrums are tall, robust perennials with upright, ribbed leaves, somewhat re-sembling corn. They grow in wet meadows and seeps, and are sometimes consid-ered a pest in pastures because of their toxicity to browsing livestock. Four spe-cies of *Veratrum* occur in California. Only one of these, *V. fimbriatum*, is easy and reliable in low-elevation gardens. *Veratrum californicum* var. *californicum* is suited only to high-elevation gardens.

Veratrums are difficult to propagate from seed, and the rhizomes do not trans-plant easily, so they are only occasionally available, even from specialty nurseries.

Some authors have placed *Veratrum* in the Melanthiaceae, along with *Xerophyllum* and *Zigadenus*.[24]

Veratrum californicum var. californicum, California corn lily, bears numer-ous small white or greenish white flowers in a wide and mostly erect panicle at the top of three- to seven-foot leafy stems. It can be a lush and showy streamside plant if the large, ribbed leaves are protected from pests. It is native to high-elevation wet places at the edges of coniferous forests or in meadows and seeps throughout much of California to Washington, Montana, Colorado, and Mexico.

Veratrum fimbriatum, fringed false hellebore, is native to wet meadows in coastal scrub in Mendocino and Sonoma Counties. Leaves are glabrous or sparsely hairy, and erect inflorescences bear white flowers that are conspicuously fringed on the margins. Flowers appear in late summer or fall. This species is on the CNPS "watch" list (List 4). Seeds are occasionally available from botanic garden plant sales and specialty seed suppliers.

Xerophyllum tenax, beargrass

Prefers: mid- to high elevation, cool, dry, sun, good drainage, acid soils
Blooms: summer

Beargrass is found at mid- to high elevations on dry, open slopes and ridges or in openings in montane evergreen forests or chaparral in the Klamath, North Coast, and Cascade Ranges, the northern Sierra Nevada, and the San Francisco Bay Area to British Columbia, Montana, and Wyoming.

A sun-loving perennial herb that grows from a tuberlike rhizome, beargrass forms dense clumps or tussocks of grasslike evergreen leaves. It thrives in well-drained, frequently shallow or rocky soils, including serpentine and volcanic ash, but dislikes dense clay and sodic or saline soils. Out of bloom it is a dense, eighteen-inch to three-foot-tall fountain of narrow, grassy foliage, green above and blue-green or gray-green beneath. In summer a leafy stem may appear, rising to as much as six feet and topped by a torchlike cluster of small, creamy white flowers.

In the wild, beargrass tends to bloom in cycles, especially in years following fire, when whole colonies bloom at once, sometimes after years of quiescence. After fruits set, plants that have bloomed die and are replaced by vegetative off-shoots from the rhizomes.

Beargrass is a handsome foliage plant, either individually or massed, particularly on banks or hillsides where the bright undersides of the leaves may be viewed from below. Unfortunately, it is difficult to propagate by seed (cold stratification may help), and it is often impossible to reestablish divisions dug from the ground. Beargrass is considered long-lived because, where its cultural requirements are met, it continues to multiply by offshoots of the rhizome, but it is not easy to keep long term in gardens.

Some authors have placed *Xerophyllum* in the Melanthiaceae, along with *Veratrum* and *Zigadenus*.[25]

Zigadenus fremontii, Fremont's zigadene, star lily

Prefers: low to mid-elevation, sun, dry, good drainage
Blooms: late winter, early spring

Of the seven species of *Zigadenus* native to California, *Z. fremontii* is perhaps the best for gardens, and it bears the largest and showiest flowers. Among the earliest bulbs to bloom in California, Fremont's zigadene is dormant by early summer. Starting in fall or early winter, it reappears on slopes in meadows, woodlands, and sagebrush scrub of coastal and foothill regions where soap plants (*Chlorogalum* spp.) thrive. It is easy to grow in gardens.

Zigadenus fremontii

Most *Zigadenus* species are toxic to both livestock and humans. Native Americans ate the bulbs of *Camassia*, with which Fremont's zigadene grows in the wild, and they weeded the camassia fields to remove these plants.

Fremont's zigadene has conspicuous greenish white flowers with bright yellow stamens and a yellow or green gland at the base of the petals. The star-shaped flowers form a rounded raceme about three inches wide and make a good early cut flower. The V-shaped strap leaves are yellow-green. Foliage clumps may be long and lank or compact and erect; in grasslands they may form a spike, while in brush they may be rangier. There is great variation among plants of this species, some varieties growing only a few inches tall instead of the more typical three feet.

Propagation is easy by seed or by division. Plants can form a beautiful patch in well-drained clay soils and full sun.

Some authors have placed *Zigadenus* in the Melanthiaceae, along with *Veratrum* and *Xerophyllum*.[26]

CALOCHORTUS

Calochortus means beautiful grass, and the name is surely appropriate. These plants traditionally have been considered members of the Liliaceae, or lily family, but some authors now place them in their own family, the Calochortaceae,[27] with one genus and about sixty species native from British Columbia south to Guatemala and east to Nebraska. Others believe that the calochortus family should be divided into several genera.

Most calochortus are native to California, and almost half of these are rare, endangered, threatened, or in decline within the state.[28] Many plants of the genus are notoriously difficult to cultivate, primarily because of their susceptibility to rot. A few, such as white globe lily (*Calochortus albus*), are commonly cultivated.[29]

The genus *Calochortus* includes the relatively familiar tall mariposa lilies and an assortment of lower-growing plants commonly known as globe lilies or fairy lanterns, cat's ears, and star tulips. Although the tall mariposas are the showiest in flower, some of the others are more tolerant of shade and water and generally more adaptable to garden conditions.

All calochortus grow from a solid bulb covered with a thin skin or with a coarse, fibrous coat. Particularly among the mariposas, a bulblet is produced within the axil of the lowest leaf at or near ground level. Thin, grasslike leaves emerge early in spring, usually dying down about the time that flowers appear. In some mariposas the basal leaf is broad, and in some it may exceed the length of the flower stem.

Regardless of the length or width, the basal leaf typically lies flat on the ground. Most mariposas are branched, and when growing vigorously they may bear several flowers on each stem. In arid, rocky situations plants may be only a few inches tall, topped by a single flower.

Flowers of the tall mariposas are shaped like an open, upright-facing bell, somewhat like a tulip but with shorter and broader petals. Flowers of the smaller globe lilies are nodding, with petals that enclose their inner parts. Star tulips have upright, bowl- or tulip-shaped flowers.

Tall mariposa flowers consist of three broad petals and three narrow, pointed sepals, which often roll back like a scroll. The fan-shaped petals are wide at the outer edge, and in some species they curve back to form a broad rim. In other species the petals stand erect, creating a vase-shaped flower. Petal texture is thick and smooth as if composed of laminated layers. Flower color ranges from white or yellow to lavender, lilac, dark rose, wine red, deep red, or vermilion. Within the flower bowl each species has a distinguishing gland—a depressed or raised area that varies in size and shape. Designs within the floral bowl may consist of only a few faint lines or flushes of contrasting colors or there may be vivid pencilings, eyespots, and tintings in bright contrasting colors enhanced by a scattering of silky hairs.

The fanciful designs and color combinations, as well as the broad forms of the flowers and their graceful movement in the breeze, led early Spanish travelers to call these lovely plants *mariposa*, or butterfly. Seed capsules are upright, slender, three-sectioned, and often beaklike in profile, containing flat, irregular seeds.

Growing Calochortus

Calochortus are propagated from seed, which usually is produced in generous quantities. Fall sowing is best, as it

Calochortus luteus

Seed pods and seed

follows the natural timing of seed germination with the soaking rains of early winter. Some growers prefer spring sowing in a greenhouse, where light and moisture can be controlled.

A standard porous potting mix can be used. Containers may be deep pots, cold frames, or outdoor seedbeds screened to keep out birds and animals. Seeds should be covered lightly with a thin layer of coarse sand or fine gravel. The sand or gravel helps to prevent water from collecting on the surface, which can cause damping off. It also conserves moisture during the dry period and provides insulation from summer heat.

The first year in the life of the seedling is most critical, as it is then that seedlings are vulnerable to damping off or drying out. During this period soil should be kept evenly moist, and seedling bulbs kept growing as long as possible. Only a single spear-leaf and a tiny, grain-sized bulblet will be produced the first year. In succeeding years more spears will appear, and by the third year the bulb will have enlarged and formed a thin coating. A few flowers may appear in the third season.

Yellowing of the foliage, which typically occurs in mid-summer, signals the beginning of the dormant period, when watering should be gradually reduced and then withheld. Occasionally seedlings will continue to grow and leaves will remain green, in which case it is best to continue to water sparingly.

Nurseries usually sell calochortus as bulbs raised from seed. These are often field grown, which tends to make them easier to maintain in the open garden. Bulbs are planted two or three inches deep in fall or early winter; this allows roots to develop but holds back development of foliage until spring.

In the garden the tall mariposas require full sun, while the light needs of the other types vary from one species to another. They all require porous, well-drained soil and good air circulation. Bulbs should be planted four to five inches deep and only a few inches apart for maximum effect. They may be protected from gophers and other burrowing animals with a wire planting basket.

Calochortus bulbs can be planted any time after dormancy begins, from late summer or fall through early winter, after which they begin to lose vitality. A top dressing of leaf mold may be supplemented with a weak solution of liquid fertilizer at monthly intervals during the growing season, or a slow-release fertilizer may be applied at planting time. It usually takes about four years for flowering to begin, and even mature plants may rest for a year or more after flowering well for several years. This should not be cause for alarm, as it is the natural life pattern of many calochortus.

Tall Mariposas for the Garden

Mariposa lilies have a reputation for being difficult to grow, but with a little attention to their cultural requirements, some have proved quite adaptable. The following are most likely to accept garden conditions if their special requirements are met.

Calochortus catalinae, Catalina mariposa

Prefers: mid-elevation, sun, coastal habitats, dry summer dormancy
Blooms: early spring to mid-spring

Catalina mariposa is a slender, branched plant with a delicate and graceful bearing. White flowers are often tinted with lilac and have a purple spot at the base of each sepal and petal. The interior of the flower is smooth except for hairs around the oblong gland. This is one of the first of the tall mariposas to flower, beginning in early spring. Its natural habitat is moderately heavy soils on slopes and in valley and foothill grasslands, open chaparral, and cismontane woodland from San Luis Obispo County to San Diego County and the Channel Islands. It is locally common in the Santa Monica Mountains and particularly conspicuous after fires. It was once abundant in the Los Angeles and Pasadena areas.

   This bulb is declining in southern California because of suburban expansion into the foothills surrounding the Los Angeles Basin. Because of its uncertain future in the wild, it is on the CNPS "watch" list (List 4). Bulbs are available in the nursery trade, although they can be difficult to find.

Calochortus clavatus, club-haired mariposa, golden butterfly tulip

Prefers: low to mid-elevation, sun, dry summer dormancy, good drainage
Accepts: clay soil
Blooms: spring, early summer

Calochortus clavatus is a stout plant with stems two to three feet tall and large, open, brilliant yellow flowers with thick petals. The circular gland is surrounded by a fringed membrane, and the anthers are dark brown or maroon. The plant is native to dry, rocky slopes, chaparral, and open forest and woodland in the Sierra Nevada foothills from El Dorado County to Mariposa County and in the Central Coast Ranges from Stanislaus County to Los Angeles County. It is not easy to maintain in gardens.

Calochortus luteus, gold
nuggets, yellow mariposa

Prefers: low to mid-elevation,
heat, sun, dry summer
dormancy, good drainage

Accepts: serpentine soils, clay

Blooms: late spring

Gold nuggets has bright yellow
flowers that seem to float above
the grasses in open meadows
and fields. The flower bowl may
be marked with faint reddish
brown lines or a median spot of
reddish brown, but it is just as

Calochortus luteus. (WILLIAM FOLLETTE)

often unmarked. Several flowers on each stem bloom in succession. The slender
stems are one foot to two feet tall and seldom branched.

Adaptable and easy to grow in gardens, gold nuggets is effective with other
native bulbs and perennials, and it is especially striking with the powder-blue flowers of *Gilia capitata*. California's most widespread mariposa, it is found in the Sierra Nevada from Tehama County to Kern County and in the Coast Ranges from
Mendocino County to Santa Barbara County. It also is native to Santa Cruz Island. Its natural habitat is heavy soils on grassy slopes and in meadows, foothill
woodlands, and open forests.

'Golden Orb', a selection raised as a field crop in Holland, has golden yellow
or bright canary yellow flowers with a central brown eyespot.

Calochortus macrocarpus, sagebrush mariposa, green band mariposa

Prefers: mid-elevation, sun, dry summer dormancy, good drainage

Blooms: summer

Sagebrush mariposa has the largest flowers of the genus in California. The pink,
soft lilac, or light purple flowers have a green median line on each segment. The
smooth petals come to a point and are set off by long, slender sepals. This mariposa
decorates sagebrush country in scrub and coniferous woodland from northeastern
California to British Columbia, generally on sandy to volcanic soils. In the garden,
plants may flower for a year or so and then disappear, perhaps because it is difficult
to replicate their soil requirements. Those who garden where rainfall exceeds about
twenty inches a year will need to protect this plant from winter rains.

Calochortus splendens, splendid mariposa, lavender mariposa

Prefers: low to high elevation, sun, dry summer dormancy, good drainage
Blooms: late spring, early summer

Splendid mariposa is a delicate-looking plant with lavender-pink flowers with purple anthers and a small purple spot of luminous purple at the base of each petal. Thick, branched hairs surround the oval gland. Native to chaparral, foothill woodlands, and coniferous forests in the Coast Ranges from Colusa County to Baja California, this mariposa prefers full sun, summer heat, and rocky soils. It flowers from late spring to early summer.

Splendid mariposa is not easy in cultivation. **'Violet Queen'**, a field-raised Dutch strain, may be easier in gardens than the species.

Calochortus superbus, superb mariposa, yellow mariposa

Prefers: low to mid-elevation, sun, dry summer dormancy, good drainage
Accepts: serpentine
Blooms: spring, summer

Superb mariposa has large flowers with intricate markings. Several flowers on each stem bloom in succession. Flower color ranges from creamy white or yellow to rose or deep lavender with a dark brown eyespot surrounded by a zone of bright yellow. The flower bowls are smooth except for a few short hairs near the A-shaped gland.

Superb mariposa is common in sun on dry slopes and in bunchgrass meadows and open woodlands of the Sierra Nevada and the Coast Ranges from Shasta County to San Diego County. Easier than most calochortus in garden cultivation, it flowers from late spring to early summer.

Calochortus venustus, butterfly mariposa, white mariposa

Prefers: mid- to high elevation, sun or part shade, dry summer dormancy, good drainage
Accepts: serpentine
Blooms: mid-spring, early summer

Butterfly mariposa has intricately decorated flowers ranging from creamy white

Calochortus venustus. (STEVE JUNAK)

or yellow to rose, velvety dark red, wine red, or occasionally delicate pink. A dark blotch low in the flower bowl may be repeated at the wide rim and enhanced by fine lines or tinting in contrasting colors. The square gland is covered with short yellow hairs, and the slender sepals tend to curl back. One to three or more flowers are borne on each one-foot to two-foot stem. The several flowers bloom in succession.

Butterfly mariposa grows in sandy to rocky soils, sometimes in decomposed granite or serpentine, in sunny meadows and open woodlands and on steep banks in the Sierra Nevada from El Dorado County to Kern County and in the Coast Ranges from San Francisco County to Los Angeles County. Given sun and sharp drainage and allowed to go dormant in summer, it may adapt well to conditions in the garden, where it can be combined with California perennials such as penstemons, monardellas, salvias, and eriogonums.

Calochortus venustus

Calochortus vestae, Coast Range mariposa

Prefers: mid-elevation, sun, dry summer dormancy, good drainage
Accepts: part shade, moderate water
Blooms: late spring, summer

Coast Range mariposa is tall and robust with a regal bearing. Large, white flowers are accented with reddish brown at the bases of the petals, and the brownish eyespot is surrounded by yellow tinting. Plants with pale yellow or pale purple flowers are sometimes found.

Although not easy to grow, this is a satisfactory garden plant where good drainage is assured. Abundant flowering can be expected from spring to mid-summer. Plants often spread by underground offsets or bulblets, which may appear some distance from the parent plant. Coast Range mariposa grows naturally in stony clay soils in grassy meadows and open forests of the North Coast Ranges in Humboldt, Napa, and Sonoma Counties.

Calochortus weedii, Weed's mariposa, late-flowered mariposa

Prefers: low to mid-elevation, sun, dry summer dormancy, good drainage
Accepts: heavy or rocky soils
Blooms: summer

This is a striking example of a group of plants that are distinguished from other mariposas technically by their chromosome count and more obviously by fringed petal margins of their intensely colored flowers and the thick, fibrous, net-veined coating of the bulbs. Most species of this group of *Calochortus*, known as Cyclobothra, are familiar only to botanists, but *C. weedii* is sometimes grown in gardens.

The large, open, deep yellow flowers of Weed's mariposa are flecked and edged with reddish brown, with long yellow hairs surrounding the round, almost naked gland. The plant is native to dry, heavy or rocky soils in chaparral in the foothills of the Santa Ana Mountains to San Diego. It flowers once every two or three years, usually in mid- to late summer. Although it may be difficult to establish, it seems adaptable to garden conditions as long as full sun and dry summer dormancy are provided.

Globe Lilies for the Garden

The diminutive globe lilies can hardly be equaled for their delicacy and distinctive bearing. Native to woodlands and forests in California, several have proven adaptable to gardens in shade or part shade and less demanding than the tall mariposas.

Globe lilies should be planted where their dainty shapes and colors can be appreciated. They add sparkle to shady borders and wooded areas and are suitable companions for other shade-loving native plants such as disporums, heucheras, and some irises. They also can be used with non-native shade plants such as vio-

lets, low campanulas, and prim-
roses. With good drainage they
may be watered throughout the
year, and if organic materials are
added to the soil, they should
make vigorous growth.

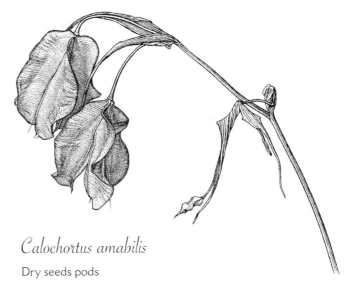

The shape of the flowers
divides the small calochortus
species into two groups. In the
first are four species of globe lilies or fairy lanterns with nodding, globe-shaped flowers. In
the second are ten species of

Calochortus amabilis
Dry seeds pods

star tulips and cat's ears, all with cup-shaped flowers. Plants in both groups are of
low stature, generally from a few inches to a foot tall, sometimes branched, and
with a long, glossy basal leaf that may rise above the flower stem but more typically lies flat on the ground. Dainty flowers occur in pearly white and several shades
of yellow, as well as lavender, lilac, and rose. Dark, irregularly shaped seeds develop in large, nodding, three-sectioned capsules.

Calochortus albus, white fairy lantern, white globe lily

Prefers: low to mid-elevation, dry summer dormancy, part shade, rocky soils,
good drainage
Blooms: spring, early summer

White fairy lantern has nodding, globe-shaped flowers composed of three overlapping petals and three shorter sepals, which may be tinted wine red, pink, pur-

Calochortus albus. (MARLIN HARMS)

plish, or brown. Plants are stout, one foot
to two feet tall, with a long, reclining
basal leaf and pointed, leaflike bracts
along the stem. A woodland plant, white
fairy lantern occurs in chaparral and open
woodlands in the Sierra Nevada foothills
from Butte County to Madera County
and in the Coast Ranges from Contra
Costa County south to the Channel Islands.

This refined plant has inspired much

admiration, and it is gradually becoming popular in gardens. Adaptable and not difficult in cultivation, it is ideal for filtered shade and elegant against a backdrop of ferns or alum root (*Heuchera micrantha*) and other shade-loving plants. A form found from the Santa Cruz Mountains to San Luis Obispo, and sometimes offered as **Calochortus albus var. rubellus**, bears rosy brown flowers from spring to early summer.

Calochortus amabilis,
golden globe lily,
golden fairy
lantern
Prefers: low to mid-
elevation, dry summer
dormancy, part shade, good drainage
Blooms: spring, early summer
Golden globe lily is a stout, branched plant with bright yellow flowers that are triangular in outline. The fringed petals are surrounded by pointed sepals. An enchanting effect is achieved by combining this globe lily with the violet-blue-flowered *Iris macrosiphon*. High shade, humusy soil, and low to moderate water generally meet its needs, although full sun in gravelly soil with little organic matter works well in cooler locations. Golden globe lily can be planted in shady dry borders, as well as in rock gardens and raised beds. It blooms in spring or early summer.

Golden globe lily is found in loamy soils and on brushy slopes or in open woodlands in the North Coast Ranges from Solano and Marin Counties to Humboldt County. In Colusa County it grows in hot openings in chaparral on serpentine soil.

Calochortus amabilis

Calochortus amoenus, rose fairy lantern, purple fairy lantern
Prefers: low to mid-elevation, dry summer dormancy, shade or part sun, good drainage
Blooms: spring, early summer

Rose fairy lantern is similar to the plant often offered as *Calochortus albus* var. *rubellus*, but with narrowly campanulate flowers that are truly rose pink. The conspicuous gland resembles a blister. When growing vigorously, the plant is much branched and bears many nodding to partially erect flowers in spring or early summer.

Rose fairy lantern is native to grassy fields and light woodlands, especially on slopes, in the Sierra Nevada foothills from Madera County to Kern County. In the wild it combines beautifully with the lavender-blue flowers of Munz's iris (*Iris munzii*). In the garden it prefers part shade and good drainage.

Star Tulips and Cat's Ears for the Garden

Star tulips and cat's ears have a sprightly charm in the perfection of their cup-shaped flowers, most of which have some delicate interior design. They are best grown with other small plants and low perennials and must be protected from rampant spreaders. They can also be grown in pots in a friable, well-drained soil mix. Plants should be allowed to dry out after flowering.

Calochortus coeruleus, beavertail-grass, blue star tulip
Prefers: mid-elevation, sun to part shade, dry summer dormancy, good drainage
Blooms: spring, early summer

Beavertail-grass has heavily bearded lilac to blue, cup-shaped flowers. A colony of this charming plant is effective in a raised border or among rocks in filtered light with other mat-forming plants. It is not easy to grow in gardens, but in sun to high shade and with moderate water and gritty soil it may establish itself and even spread gradually by volunteer seedlings.

Beavertail-grass is native to gravelly soils in coniferous forests, open woodlands, and grassy slopes of the Sierra Nevada from Lassen and Tehama Counties to Amador County.

Calochortus monophyllus, Sierra star tulip, yellow star tulip
Prefers: mid-elevation, part shade, dry summer dormancy, good drainage
Accepts: full sun near coast
Blooms: spring, summer

Sierra star tulip bears attractive, upright, star-shaped, bright yellow flowers with a brown spot at the base of each petal. It grows in clayey loam soils on wooded slopes in the foothills of the Sierra Nevada from Shasta County to Tuolumne County. In the garden, with little attention, Sierra star tulip may spread by volunteer seedlings. Low growing, it is particularly effective planted in groups with *Iris tenuissima* or mats of *Monardella odoratissima* or *M. purpurea*.

Calochortus nudus, naked mariposa

Prefers: mid-elevation, some moisture, cool temperatures, light shade, good drainage
Blooms: summer

Naked mariposa has small, white to pale lavender, cup-shaped flowers, smooth inside except for a fringed membrane below the transverse gland. This refined plant is best in a cool rock garden or a woodsy border. It needs protection from rampantly spreading plants. It is native to moist meadows and seeps in the Klamath and Cascade Ranges to southwest Oregon.

Calochortus tolmiei, Tolmie's star tulip, pussy ears

Prefers: low to mid-elevation, light shade, dry summer dormancy, good drainage
Accepts: full sun near coast
Blooms: spring, summer

Calochortus tolmiei

Tolmie's star tulip or pussy ears is about four to six inches tall with white flowers delicately tinged with lavender and silken hairs on the inner surface. The glossy basal leaf may be twice as long as the flower stem. Plants should be grown in the filtered shade of a woodland garden, as border edging with other low plants, or in a rock garden.

Tolmie's star tulip is native to dry, grassy slopes, coastal bluffs, and open woodlands or redwood forests of the North Coast Ranges south to Santa Cruz County and north to Washington. Although not easy to grow in gardens, where it is established it will persist and even spread by volunteer seedlings.

Calochortus umbellatus, Oakland star tulip

Prefers: low elevation, part shade, dry summer dormancy, serpentine
Accepts: full sun near coast
Blooms: early spring

This dainty star tulip is native to dry, brushy slopes and open grassland in the San Francisco Bay Area, generally on serpentine soils. The plant is only about four to six inches tall, and the flowers are white with purple markings near the base. It should be displayed where its elfin charms will not be overwhelmed by other plants. Because of its limited distribution in the wild, Oakland star tulip is on the CNPS "watch" list (List 4). Seeds are occasionally available from plant society and botanic garden sales.

Calochortus uniflorus, pink star tulip, large-flowered star tulip

Prefers: low elevation, coastal and vernally moist habitats, sun, dry summer dormancy
Accepts: high shade, some summer moisture
Blooms: spring

Pink star tulip has pink to pale lavender flowers with a scattering of hairs below the gland and sometimes a purple spot at the base. An inhabitant of seasonally wet places, such as vernal pools and marshy meadows, it is also found in scrublands and coniferous forests from coastal Monterey County to Oregon.

Calochortus uniflorus. (ROBERT CASE)

Calochortus uniflorus

   This is an adaptable garden plant, accepting full sun to high shade and tolerant of some water throughout the year, although it prefers dryness when dormant. It increases by underground offsets and soon forms extensive colonies. It is
effective with low plants of similar requirements such as blue- or purple-flowered
penstemons.

TRILLIUMS

Thirty to forty species of *Trillium* occur in North America and Asia. Only five are
found in California, ranging in the north half of the state throughout the Coast
Ranges, at mid-elevations in the Sierra Nevada, and across the Siskiyou Mountains. Found in humus-rich soils that are moist in spring, trilliums prefer the dappled
light of the woodland or forest floor, although they occasionally grow beneath
coastal scrub or coyote brush and often in grassland among scattered trees.

Trilliums traditionally have been considered members of the broadly defined lily family,[30] but more recently have been placed in their own family, the Trilliaceae.[31] Some authors consider them to be only distantly related to the lilies.[32]

Arising from short, thick rootstocks, these beautiful and distinctive perennials display a three-petaled flower over a whorl of three broad leaves. In some species the leaves are mottled in contrasting maroon or subtle silver-green. Flower color most often ranges from white to pink to deep maroon, although flowers of some species are brownish, greenish, or chartreuse. Plants range from three to eighteen inches tall, depending on the species. Older plants may have numerous stalks, each bearing three leaves and a single flower, which combine to form large, dramatic clumps.

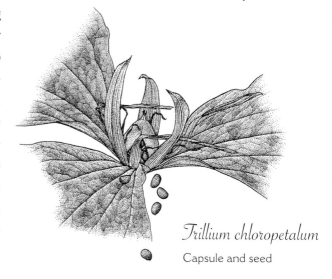

Trillium chloropetalum
Capsule and seed

Growing Trilliums

California trilliums vary in their adaptability to garden conditions. If provided with well-drained soil and placed in part shade, most are relatively easy to grow and eventually will form sizable clumps. Depending on the species, some trilliums make a bold statement, while others add a quiet charm in the woodland or rock garden. Specialty nurseries and a few retail nurseries offer them as rootstocks in fall or in pots at other times of the year.

Trilliums can be propagated by seed or by division of rootstocks, both of which methods are slow. With the exception of *Trillium rivale*, seed requires two seasons of cold before fully

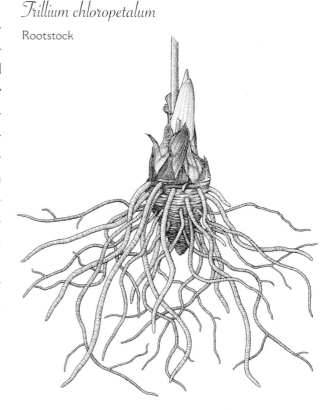

Trillium chloropetalum
Rootstock

germinating. After this it may be three to seven years before plants bloom. Plants spread slowly, and once a sizable clump has been established, divisions should be made only every two or three years.

Growers sometimes increase stock by cloning. This procedure takes five years to produce a blooming plant and involves making small incisions at the end of the rootstock, which induces the formation of new rootstocks. After two years these are cut from the parent plant and grown on to produce a flowering plant.

Trilliums for the Garden

California trilliums are divided into two groups, based on the manner in which the flowers are presented. Trilliums in one group (*Trillium albidum*, *T. angustipetalum*, and *T. chloropetalum*) have sessile flowers; that is, the flowers come directly from the whorl of three leaves, without any stem. Those in the other group (*T. ovatum* and *T. rivale*) display flowers held above the leaves by a short stem.

Trillium albidum, with white flowers with short, broad petals, is similar to *T. chloropetalum* and has similar uses.

Trillium angustipetalum, with narrow, elongate, plum-purple flowers, is similar to *T. chloropetalum* and has similar uses.

Trillium chloropetalum, common trillium, giant trillium
Prefers: low to mid-elevation, part shade, some summer moisture
Blooms: early spring
Common trillium is ten to eighteen inches tall, with bold leaves that often exhibit striking mottling in shades of deep green or maroon. The flowers, which are large and sessile,

Trillium chloropetalum. (CHARLES KENNARD)

range from white, pink, or red through purple-brown, amber, and green to almost yellow. Some have thin petals, while others have large, wide petals with a waxy texture. With protection from the hottest sun and a bit of summer moisture, the plant will settle down and begin its steady progression to a bold and handsome clump.

Common trillium is found in coastal scrub and coastal and interior open forests and woodlands in the North Coast Ranges and the San Francisco Bay Area, usually on moist banks and slopes in alluvial soils. It prefers occasional summer water

Trillium chloropetalum

but tolerates some summer dryness, and in northern parts of the state it may persist without irrigation along the coast or inland on a cool, shady, north-facing slope.

Trillium ovatum, western trillium, Pacific trillium
Prefers: low to mid-elevation, coastal habitats, cool, shade, moisture, good drainage
Blooms: early spring
Western trillium bears large, wide-spreading, showy white flowers, often fading to pink, that are poised in a slight nod. A short stem holds the flower a few inches above the three elegantly tapered, broad leaves. Plants are six to fifteen inches tall. The most widespread of the native trilliums, western trillium is found along the coast of northern California and in both coastal and interior

Trillium ovatum

ranges of Oregon, Washington, and British Columbia. It usually grows in deep shade on moist slopes in coniferous forests.

Within its natural range, where appropriate conditions are provided, western trillium can be successfully grown in gardens. However, it is quite intolerant of warm interior climates, and if it does not fail, it sulks, sending up a feeble whorl of leaves year after year that shrivel early without producing a flower. Even under ideal conditions, plants are slow to increase in the garden. The best chance for success with this plant is to place it in a cool, shady location and provide some summer moisture.

There are two double-flowered forms of western trillium in cultivation, both of which are as rare in the wild as they are beautiful.

Trillium rivale (Pseudotrillium rivale),
brook wakerobin

Prefers: low to mid-elevation, shade, moisture, good drainage

Blooms: early spring

Trillium rivale

Brook wakerobin is a delightful elfin trillium, seldom over four inches tall, with one-inch-wide flowers held slightly above small, rounded leaves with pointed tips. The leaves are sometimes etched with silver-green. The spreading petals are white to pale pink and often spotted with purple.

This plant is restricted to the extreme northwest corner of the state in Del Norte and Siskiyou Counties into southwestern Oregon. It is found on stony, shaded slopes in yellow pine forests, often near rocky streambanks, and often on serpentine soils.

Brook wakerobin requires a shady spot in the garden and gritty soil that stays damp, but not wet, at all times. Because of its diminutive size and delicate beauty, it should be displayed where it can be seen close up. It makes a good container plant or rock garden subject and will multiply quite rapidly if its cultural requirements are met.

ALLIUMS AND BRODIAEAS

Alliums and brodiaeas (including the genera *Brodiaea*, *Dichelostemma*, and *Triteleia*) are perennial herbs with linear leaves and flowers in dense or open umbels or umbel-like inflorescences. Native to many regions in California, they bring spring and sometimes summer color to a variety of habitats, including grassy meadows, open woodlands, and chaparral.

After flowering, the plants wither and die back to the ground, remaining dormant until the following year. There are exceptions, but most of these cheerful plants are easy to grow if given good to excellent drainage and full to part sun, making them fine candidates for a low-maintenance rock garden.

ALLIUMS

California is famous for its alliums, or wild onions, which form carpets of vivid color over gravelly inclines and dry fields. Gardeners are familiar with the culinary uses of alliums, for the group includes onion, garlic, chives, shallot, and leek. The genus *Allium* is known for the strong onion odor that emanates from all parts of the plants when crushed. Alliums are perennial herbs with flowers that appear to arise from the end of a leafless stem.

The genus *Allium* is highly developed in California, with about forty species and many subspecies occupying habitats from deserts to sea bluffs to alpine bogs. All are at least interesting, and some are quite showy.

Unlike brodiaeas, which grow from corms, alliums grow from bulbs made up of tightly packed, fleshy scales surrounded by a thin, dry coat called a tunic. In winter or early spring each bulb produces one or more narrow, cylindrical or flattened leaves, followed by one or more naked stalks, each topped with an umbel of small, star-shaped or bell-shaped blossoms.

Among California alliums the range of flower color is narrower than for the genus worldwide—from reddish purple through pink to rose and, most commonly, white. All aboveground portions of the plants wither after flowering, but the flowers often retain their shape and color and are attractive in dried arrangements.

Growing Alliums

A few California alliums are large and prolific enough to create substantial drifts in open areas of the garden or in perennial borders, but most are better suited to rock gardens or container plantings. Most are easy to grow. Those that in nature grow on sun-baked, gravelly slopes should be allowed to dry out thoroughly in summer. Those from higher mountain seeps and bogs should be kept moist in spring and summer months. Some, such as *Allium unifolium*, tolerate a range of garden conditions. A well-drained soil or potting mix is desirable for all alliums and essential for those that are at home in gravelly screes.

Most alliums grow and bloom best in full or nearly full sun. They are troubled by few pests or diseases, and almost all are easily propagated either by fall-sown seed or by division of bulbs.

Alliums for the Garden

Many alliums can be grown successfully in gardens. Among them are the following species.

Allium acuminatum, Hooker's onion, taper-tipped onion
Prefers: low to mid-elevation, sun, dry summer dormancy, good drainage
Blooms: spring, early summer
Hooker's onion is a striking plant with round heads of bright rose-pink or reddish purple flowers. It grows naturally in loose, stony soils in yellow pine forests and grassy openings in chaparral from the North Coast and Cascade Ranges north to British Columbia and east and south to the Rocky Mountains and Arizona.

Flowers appear from spring through early summer on scapes four to twelve inches tall. Plants prefer dryness in summer. Suitable companions are other drought-tolerant bulbs and perennials such as monardellas, eriogonums, or penstemons. Hooker's onion is a good choice for rock gardens or for naturalizing on warm slopes among other low-growing plants. It can be difficult to establish in gardens.

Allium amplectens, narrow-leaved onion
Prefers: low to mid-elevation, sun, dry summer dormancy
Accepts: serpentine
Blooms: late spring, early summer
This onion has narrow leaves and rounded umbels of white or pink flowers that fade to creamy silver. It seeds heavily, so seed capsules should be removed if many volunteers are not desired. Narrow-leaved onion is native to clay soils or serpentine on sunny slopes in openings in coniferous forests and foothill woodlands of central and northwestern California to British Columbia.

Allium cratericola, Cascade onion, crater onion
Prefers: mid-elevation, sun, dry summer dormancy, excellent drainage
Blooms: late spring, summer

Cascade onion is a variable species with pale pink to rose-pink flowers on short stems and one or two flat, straight leaves. This onion is native to chaparral and foothill woodland on granitic soils, open volcanic rubble, or serpentine scree in the Klamath and North Coast Ranges and the Sierra Nevada in northern California to the western Transverse Ranges and the San Jacinto Mountains in southwestern California. Seeds are sometimes available from plant societies or specialty nurseries.

Allium crispum, crinkled onion

Prefers: low to mid-elevation, sun, dry summer dormancy
Blooms: spring

Crinkled onion has open umbels of lovely red-purple flowers on eight- to twelve-inch scapes. The three inner petals have crinkled edges, giving the plant its common name. Native to clay or serpentine soils on slopes in valley grasslands and open woodlands in central western California, it is easy to grow in gardens if given a spot where it can be kept dry in summer.

Allium dichlamydeum, coast onion

Prefers: low elevation, coastal habitats, sun, dry summer dormancy, good
   drainage
Blooms: spring, summer

Coast onion has flowers of deep rose-purple topping stout scapes about ten inches tall. Native to coastal prairie and coastal scrub along the north and central coast

from Mendocino to Monterey County, it thrives in dry, often rocky soils and sometimes is seen in thick colonies on sheer rock faces.

The showy flowers appear in late spring or summer. The petals hold their color until seeds set, extending the effect of the flowering season. Given sun and well-drained soil, this allium is quite adaptable to garden conditions.

Allium dichlamydeum. (ROBERT CASE)

Allium falcifolium, sickle-leaf onion

Prefers: low to mid-elevation, sun, dry summer dormancy, excellent drainage

Blooms: spring, summer

Sickle-leaf onion is found on dry, rocky soils, usually on serpentine scree or out-crops, in the San Francisco Bay Area and the North Coast Ranges from Santa Cruz County to southwestern Oregon. Stems and leaves are thick and flat, and rounded clusters of flowers appear in spring or summer. Low-elevation plants of this species have intensely wine-red flowers, while most higher-elevation plants are lighter in color. Sickle-leaf onion is an especially fine candidate for sunny rock gardens, but it can be difficult to establish in cultivation.

Allium falcifolium

Allium haematochiton, red-skinned onion

Prefers: low to mid-elevation, sun to part shade, dry summer dormancy, good drainage

Blooms: summer

Easy to grow, this dwarf allium has dark green leaves and compact umbels of at-tractive bi-colored flowers, white to rose with darker midveins. It is common on dry slopes, ridges, coastal sage scrub, and grassland in the southern Coast Ranges, Peninsular Ranges, and western Transverse Ranges to northern Baja. It multiplies slowly, eventually making a nice clump.

Allium hyalinum, glassy onion
Prefers: low to mid-elevation, sun, dry summer dormancy, good drainage
Accepts: some summer water
Blooms: spring
This is an attractive plant about a foot tall with large, open umbels of white to pink-tinged flowers with a sparkling, transparent appearance. Flowers may appear in early spring and continue into mid-spring. Native to valley grassland and foothill woodland on rocky slopes in the Sierra Nevada foothills and the San Joaquin Valley, this adaptable allium will accept summer irrigation in well-drained soil. Good companions are early-flowering brodiaeas with a groundcover of five-spot (*Nemophila maculata*), or baby blue-eyes (*N. menziesii*). *Allium hyalinum* can seed heavily when it dies down for the summer.

Allium peninsulare, Mexicali onion
Prefers: low to mid-elevation, part shade, dry summer dormancy, excellent drainage
Accepts: full sun near coast
Blooms: spring, late spring
Mexicali onion is native to dry slopes and flats in the Sierra Nevada foothills, the Sacramento Valley, the San Francisco Bay Area, and the South Coast Ranges to southern Oregon and northern Baja California. It has wine-red flowers and needs excellent drainage and, except near the coast, part shade or a north-facing slope.

Allium platycaule, broad-stemmed onion, pink star onion
Prefers: mid- to high elevation, sun, dry summer dormancy, rocky soils, excellent drainage
Blooms: spring, summer
Broad-stemmed onion grows in sagebrush scrub and forest openings on gravelly slopes and ridges in the Sierra Nevada to southern Oregon and western Nevada. It prefers sunny scree conditions in the garden. A showy plant, it bears large, lax heads of rose-pink flowers atop a flat, eight-inch stem from early spring into summer. This onion prefers higher elevations but adapts to low-elevation gardens if given excellent drainage and possibly also some summer irrigation.

Allium praecox, early onion
Prefers: low elevation, part shade, dry summer dormancy, good drainage
Blooms: late winter, early spring

Early onion grows on shaded slopes and in canyons in chaparral or oak woodland in southern California and northern Baja California. This is one of the first native onions to bloom, usually in late winter or early spring. A stout ten-inch scape bears an open umbel of rose-pink, pale pink, or white flowers. In the garden it should be grown in dry shade and loose soil, but even under the best conditions it increases slowly and can be difficult to maintain in gardens.

Allium serra, jeweled onion
Prefers: low to mid-elevation, sun, dry summer dormancy, excellent drainage
Blooms: spring, early summer
Jeweled onion forms splashes of vivid color in grassy fields, on dry slopes and rock faces, and in serpentine soils in the North Coast Ranges, the San Francisco Bay Area, and the interior South Coast Ranges. The rose-pink flowers appear in tight umbels on eight- to twelve-inch scapes. Attractive and easy to grow, especially in gritty soil in a rock garden or raised bed, it spreads gradually with no attention from the gardener. It needs a dry summer dormancy. Plants are sometimes sold as *A. serratum*.

Allium siskiyouense, Siskiyou onion
Prefers: mid- to high elevation, sun or part shade, excellent drainage
Accepts: some summer water
Blooms: late spring, early summer
Siskiyou onion is found on rocky slopes, often on serpentine soils, in the Klamath and North Coast Ranges to southwestern Oregon. The flowers are rose with darker midveins or white to pale rose, aging to rose red or red-violet. This low-growing onion needs excellent drainage and may do better in the garden with some summer water. Difficult to maintain in the open ground, it may do best in containers or in the front of a rock garden.

Because of its limited distribution, it is on the CNPS "watch" list (List 4). Seeds are occasionally available from specialty suppliers and seed exchanges.

Allium unifolium, one-leaf onion
Prefers: low to mid-elevation, coastal habitats, sun, dry summer dormancy
Accepts: some summer water
Blooms: spring, early summer
One-leaf onion is among the larger wild onions and surely the most ornamental, resembling a stocky brodiaea. It is also one of the more prolific, offering ever more

lavish displays from one year to the next, and it accepts almost any garden conditions except deep shade. It can even become a garden pest.

This onion is native to chaparral, coniferous and mixed evergreen forest, and
grassy streambanks in northwestern California and southwestern Oregon. Each

Allium unifolium

bulb produces flattened blue-green leaves and often multiple flower stems with umbels of large, showy, rose-pink to white flowers. With water and continued mild temperatures, the show may be prolonged from mid-spring well into summer. *Triteleia laxa*, with lively blue-purple flowers, makes an especially fine companion.

Allium unifolium. (SAXON HOLT)

Allium validum, Pacific mountain onion, swamp onion
Prefers: mid- to high elevation, sun, wet places, summer water
Blooms: summer

Pacific mountain onion has broad, shiny green leaves with scapes two to three feet tall bearing large clusters of pink to creamy white blossoms. Native to montane coniferous forests and wet places in open meadows in the Sierra Nevada and Cascade Ranges to British Columbia, Idaho, and Nevada, this allium needs plenty of summer water and thrives in year-round seeps. It is difficult to establish in low-elevation gardens.

BRODIAEAS

Brodiaea is the common name for a variable group of plants that includes three genera—*Brodiaea*, *Dichelostemma*, and *Triteleia*—and about forty species. Brodiaeas usually are placed with alliums in the Alliaceae (and thus formerly with the lilies), but recent research has suggested that brodiaeas should be placed in their own family, the Themidaceae, and may be more closely related to the hyacinths than to the alliums.[33]

Unlike the alliums, the leaves of brodiaeas do not have an onion odor and the plants grow from corms, not true bulbs. Like the alliums, all bear flowers in umbels in which many individual flower stalks radiate from a common point. Brodiaeas bloom from early spring into early summer, followed by summer dormancy. They are concentrated in northern California, but extend south to northern Baja California, north to British Columbia, and east to the Rocky Mountains.

Plants of the genus *Bloomeria*, or goldenstars, are similar to brodiaeas in that they grow from corms and bear flowers in umbels, but the flower stalks, or pedicels, are jointed and the filaments are basally cuplike. In general appearance, they are most similar to the triteleias. Goldenstars are found in grassland, open woodland, and chaparral from central to southern California and northern Baja California.

Brodiaeas are a conspicuous element of the California landscape in spring. Some cover many square miles of grassland, open woodland, rocky hillside, and chaparral with a haze of blue, lavender, or white from early spring through early summer. Higher-elevation rocky slopes and montane meadows are frosted with drifts of their flowers in summer.

In mild climates brodiaeas appear in early winter to mid-winter, when their grassy basal leaves emerge from buried corms. At this time the roots are growing rapidly, and corms are multiplying by basal or stoloniferous offsets. In undisturbed sites self-sown seedlings and new corms may produce a stand dense enough to give the effect of turf. Some brodiaeas produce starlike basal leaf rosettes from mature corms.

Later in spring, often after the foliage has yellowed, a naked scape rises from the rosette bearing an elegant umbel-like arrangement of flowers. The height of the scape, length of the pedicels, size, shape, and color of the flowers, and time of flowering vary with climate, growing conditions, and genetics. Seed sets within a month or so after flowering, when the foliage has died back. Corms remain dormant until fall.

Growing Brodiaeas

Most brodiaeas are easy to grow. In containers, use a medium to moderately heavy soil mix that drains well. Occasional feeding with a mild fertilizer in fall or winter will stimulate root and leaf growth. Withhold fertilizer when flower scapes appear and water sparingly. Watering usually can be safely halted when flowers open.

After flowering, seed heads should be removed unless seed is desired. Cut plants back and store containers in a cool, dry place until fall, when watering and fertilizing should be resumed. Every few years corms can be lifted, divided, and repotted or given to friends.

In the open ground brodiaeas seem adapted to almost any soil. Where gophers or other burrowing predators are a problem, corms may be started in sunken wire cages or in pots. Once established, disturbance of the ground and spreading of cormlets by animals, even gophers, may promote rather than inhibit proliferation. Summer dryness is ideal, but some brodiaeas tolerate occasional summer watering, especially if they are protected from hot sun. A few accept quite frequent summer watering.

Triteleia laxa

Corm, corm with cormlets

Growing from seed is usually the best way to propagate brodiaeas. Seeds germinate readily during winter, but several months of cold stratification may be needed if seeds are not fresh. Seedlings prefer cool, but not cold, conditions, although they can tolerate occasional mild freezing for short periods. They require good air circulation. Seedling growth can be hastened with a mild liquid fertilizer, and watering should be continued until foliage begins to turn yellow. Seed pans then should be dried and stored in a cool, dry place until fall, when seedlings can be grown on for another season. The tiny corms can be removed from the pot and refrigerated in plastic bags, then repotted in fall. Most brodiaeas require at least three years from seed to reach flowering size.

Propagation from corms is not difficult. Some species of *Dichelostemma* and *Brodiaea* produce corm offsets prolifically. The cormlets can be started in pots in rich, loose soil with frequent fertilizing and annual division.

True brodiaeas grow from globose corms covered with a dark, patterned outer coat. Corms produce two to fifteen offset cormlets in the axils of old leaf bases. Cormlets do not flower until they reach a mature size, which generally takes a year or two. The leaves are rounded on the bottom and crescent-shaped in cross section. Flowers are in open umbels on a leafless stem in late winter or early spring into summer, and petals and sepals usually are glossy and waxy with a pronounced midvein.

Most plants of the genus *Brodiaea* are diminutive and ideally suited to the rock garden or the front of the dry border. They are also superb in pots. Corms may multiply rapidly in the garden. In the wild, they often grow in clay hardpan and may be submerged in winter for several months, but in the garden they benefit from normal drainage.

Brodiaeas are effective combined with spring annuals in a low-growing meadow planting. Native grasses such as Idaho fescue (*Festuca idahoensis*) or California oat grass (*Danthonia californica*) and perennials or bulbs such as blue-eyed grass

(*Sisyrinchium bellum*), California buttercup (*Ranunculus californicus*), checkerbloom (*Sidalcea malviflora*), Chinese houses (*Collinsia* spp.), cream cups (*Platystemon californicus*), and wild onions such as *Allium amplectens*, *A. hyalinum*, or *A. unifolium* will combine with the smaller brodiaeas to good effect.

Mid-height brodiaeas, such as *Brodiaea elegans*, are effective in many garden situations, but may look most natural in meadow plantings. They can be grown with taller grasses such as purple needlegrass (*Nassella pulchra*), melic grasses (*Melica* spp.), reedgrasses (*Calamagrostis ophitidis or C. rubescens*), or California fescue (*Festuca californica*). Mid-height brodiaeas also are useful as fillers in open dry shrub or perennial plantings, blending well with coyote mint (*Monardella* spp.), California fuchsia (*Epilobium* spp.), manzanita (*Arctostaphylos* spp.), and California lilac (*Ceanothus* spp.).

Brodiaeas for the Garden

Plants of the genus *Brodiaea* are restricted to western North America, from Vancouver, British Columbia, to northern Baja California. Many species are rare, including *B. orcuttii*, a species native to vernal pools in southern Riverside and San Diego Counties, and *B. pallida*, an extremely endangered brodiaea known from a single population in Tuolumne County. The species described below are more common, and corms or plants are available from plant society and botanic garden sales or from specialty seed and bulb suppliers.

Brodiaea bridgesii, see *Triteleia bridgesii*

Brodiaea californica, California brodiaea
Prefers: low to mid-elevation, sun or part shade, dry summer dormancy
Blooms: late spring, early summer
California brodiaea is the tallest and showiest of the true brodiaeas, as well as the most common. It is widespread in open grasslands, woodlands, and chaparral, often on gravelly clay soils or serpentine in the North Coast Ranges and the northern Sierra Nevada into southwestern Oregon. The waxy pale lavender or violet (rarely pink or white) flowers are borne in large umbels on stout eight- to thirty-inch scapes.

California brodiaea is easy to grow if provided a dry summer dormancy. Corms do not offset readily, so seed propagation is best.

Brodiaea congesta, see *Dichelostemma congestum*

Brodiaea coronaria, crown brodiaea

Prefers: low to mid-elevation, sun, vernally moist, dry summer dormancy
Blooms: summer

Crown brodiaea is four to ten inches tall and bears flowers ranging from soft lilac to lavender-purple, sometimes with a darker petal midvein. One of the later brodiaeas to bloom, it is common in vernally moist places and on volcanic mesas or dry slopes in valley grassland, foothill woodland, and coniferous forest on clayey or gravelly alkaline soils in the Cascade Ranges and the northern and central Sierra Nevada to British Columbia.

Brodiaea elegans, harvest brodiaea

Prefers: low to mid-elevation, heat, sun, dry or moist conditions
Accepts: heavy soil, some summer water
Blooms: summer

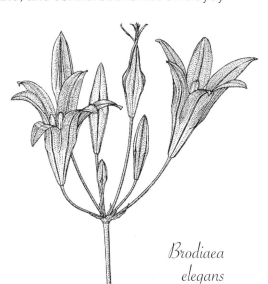

*Brodiaea
elegans*

Harvest brodiaea is one of the later of the tall brodiaeas to flower, its elegant umbels making a lovely show in summer among golden drying grasses. The lavender-blue to blue-purple flowers with yellowish centers appear almost translucent. The loose umbels are borne on strong, wiry scapes twelve to eighteen inches tall.

Native to moist or dry slopes in valley grassland or open foothill woodlands in the Klamath and North Coast Ranges and the northern Sierra Nevada to southwestern Oregon, this is a reliable and adaptable garden plant. It tolerates occasional summer water and heavy soils, and it multiplies well.

Brodiaea hyacinthina, see *Triteleia hyacinthina*

Brodiaea ida-maia, see *Dichelostemma ida-maia*

Brodiaea jolonensis, Jolon brodiaea
Prefers: low elevation, sun, dry summer dormancy
Blooms: late winter, spring
Jolon brodiaea is four to ten inches tall, with blue-lavender flowers with a deeper-colored interior. It is native to grassland and foothill woodlands on clay soils in the South Coast Ranges to Baja California.

Brodiaea laxa, see *Triteleia laxa*

Brodiaea lutea, see *Triteleia ixioides*

Brodiaea minor, dwarf brodiaea
Prefers: low elevation, sun, dry summer dormancy
Accepts: clay soils
Blooms: late winter, spring
Dwarf brodiaea is delicate in appearance and only about six inches tall with star-like flowers that usually are bluish to lilac-blue. It grows in gravelly clay soils in grasslands and vernally wet places in the Central Valley and foothill woodlands of the Sierra Nevada and northern San Joaquin Valley. It should be planted where the flowers can be appreciated up close, as in the front of a rock garden.

Brodiaea multiflorum, see *Dichelostemma multiflorum*

Brodiaea peduncularis, see *Triteleia peduncularis*

Brodiaea pulchella, see *Dichelostemma congestum*

Brodiaea purdyi, Purdy's brodiaea
Prefers: low elevation, sun, dry summer dormancy
Blooms: early to mid-summer
Brodiaea purdyi is similar and closely related to *B. minor*; it may be a subspecies.[34] It grows in open woodland and on volcanic plateaus and serpentine barrens in the foothills of the southern Cascade Ranges and the northern Sierra Nevada. The narrow flower petals are deeper lilac than those of *B. minor*, often with a dark central midvein.

Brodiaea stellaris, star-flowered brodiaea
Prefers: low to mid-elevation, coastal, sun, dry summer dormancy
Blooms: spring, late spring
Star-flowered brodiaea has petals that open flat in a starlike arrangement, and the dark midvein striping is often pronounced. It is native to forest openings, often on serpentine soils, in the North Coast Ranges.

Brodiaea terrestris, earth stars, dwarf brodiaea
Prefers: low to mid-elevation, coastal, sun, dry summer dormancy
Blooms: spring, early summer
Earth stars is one of the best of the low-growing brodiaeas for the garden and the most dwarf brodiaea found in the wild. Native to sea bluffs and coastal prairie, as well as the North Coast Ranges, Sierra Nevada foothills, and Transverse and Peninsular Ranges to southwestern Oregon, it typically is less than three inches tall, with the flower umbel often resting on the soil. The leaves form an attractive starlike rosette, and the flowers are bright, waxy, blue-lavender with paler centers and a yellowish white midvein.

This tiny brodiaea is excellent in pots.

Brodiaea terrestris

Dichelostemmas for the Garden

Plants of the genus _Dichelostemma_ grow from globose corms with dark outer corm coats that lack any particular pattern. Mature leaves have a distinct keel on the underside. Flowers generally appear in a dense umbel-like arrangement, but often are clustered into a tight head. Flowers vary from red or pink to blue-purple and lack the glossy, waxy appearance of those of *Brodiaea* species. Plants of the genus *Dichelostemma* are distributed throughout the western United States but are concentrated in northern California.

Dichelostemma includes some of the most peculiar and intriguing members of the brodiaea complex. Firecracker flower (*D. ida-maia*) is a botanic oddity, with astonishing bright red, green-tipped tubular flowers. Twining snake-lily (*D. volubile*) is a marvel of adaptive evolution, with crystalline, crinkled pink flowers and vining scapes that scramble through shrubbery seeking sunlight.

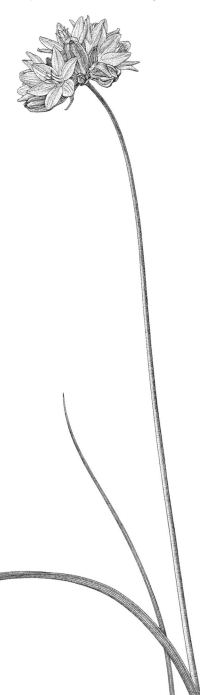

Dichelostemmas multiply quickly from cormlets, and they also are easy from seed. Most are quite tall, with scapes eighteen to over forty inches in height. On rocky or extremely poor soil, they may be half as tall. Long scapes usually benefit from support by grasses and low shrubs. Many dichelostemmas tolerate occasional summer watering.

As with plants of the genus *Brodiaea*, dichelostemmas are particularly effective in meadowlike plantings, and they also are attractive growing out of low, loose-textured shrubs, such as *Rosa spithamea* or *Berberis aquifolium* var. *repens*, and medium-height manzanitas.

Dichelostemma capitatum, blue dicks, wild hyacinth
Prefers: low to mid-elevation, heat, sun, dry, good drainage
Blooms: late winter, early spring

Dichelostemma
capitatum

Blue dicks is the most common dichelostemma, found in open woodlands, scrub, desert, and grassland habitats throughout California into Oregon, Utah, New Mexico, and northern Mexico. It may flower in late winter on warm, south-facing banks in mild climates. A prolific grower, and tolerant of almost any soil or exposure, it prefers heat, sun, and good drainage.

Blue dicks increases rapidly by offsets and is an easy plant to naturalize. In the sheer numbers of tight heads of funnel-shaped flowers on two- to three-foot stems, it can color many acres in soft blue-violet. Plants with pure white and rose-colored flowers are found, as well as an unusually large form from Santa Cruz Island in southern California.

Dichelostemma congestum, ookow
Prefers: low to mid-elevation, sun or part shade, dry, good drainage
Blooms: spring
Ookow is similar to but showier than *Dichelostemma capitatum*, with larger, later-blooming flowers. The two plants can be naturalized together for a longer season of bloom. Native to open woodland and grasslands in northwestern California to British Columbia, this plant bears rich lavender-blue flowers in congested heads on sturdy two- to three-foot stems. It makes an excellent cut flower and is effective in drifts in the middle of a dry border.

Dichelostemma ida-maia, firecracker flower
Prefers: low to mid-elevation, part shade, dry, good drainage
Blooms: spring
Firecracker flower is perhaps the most sought-after dichelostemma. This is a stately plant, up to twenty inches tall, with coarse, grasslike leaves. It produces a loose, upright umbel of buds that curve downward as they open into crinkled, crepelike tubes to two inches long. Each scarlet-red tube is tipped with projecting and recurved tepals of lime green and chartreuse.

Dichelostemma ida-maia. (ROBERT CASE)

This plant is striking in a variety of situations and combinations. It responds well to pot culture if protected from excessive wind. In the open ground it is effective with clumping grasses such as California fescue (*Festuca californica*) or the nonnative blue oat grass (*Helictotrichon sempervirens*).

Firecracker flower is native to redwood and mixed evergreen forests, or grasslands near the coast, in northwestern California from the North Coast and Klamath Ranges to southwestern Oregon.

Rose firecracker flower, sometimes called *Dichelostemma* x *venustum*, is a naturally occurring hybrid of *D. ida-maia* and *D. congestum* or *D. multiflorum*. This plant has bright magenta-pink flowers with white inner segments, each hanging on a short stalk. It is attractive, easy to grow if provided a dry summer dormancy, and prolific in corm offsets. It is often offered in the nursery trade as **D. x 'Pink Diamond'**.

Dichelostemma multiflorum, manyflower brodiaea, wild hyacinth
Prefers: low to mid-elevation, sun, dry, good drainage
Blooms: spring
This is another plant to naturalize with *Dichelostemma capitatum*, as its showy clusters of soft lilac or lavender flowers appear about a month later and will extend the season of bloom. It is native to but not common in foothill grassland, open woodland, and scrub in northwestern California, the Sierra Nevada, and the San Francisco Bay Area to southern Oregon. It thrives in sun and heat and makes an excellent cut flower.

Dichelostemma pulchellum, see *D. capitatum*

Dichelostemma volubile, twining snake-lily
Prefers: low to mid-elevation, sun or part shade, dry, good drainage
Accepts: some summer water, full sun
Blooms: late spring, early summer
Twining snake-lily grows in chaparral, scrub, and open woodland in the foothills of the interior North Coast Ranges, the Cascade Ranges, and the Sierra Nevada. Its

Dichelostemma multiflorum

sleek, reddish green flower scapes, up to five feet long, interlace sinuously along the ground or grow casually upright by coiling about the branches of nearby shrubs.

Pale rose-pink to bright pink (or occasionally white) clusters of crinkled flowers are borne on the ends of each twining scape in late spring to early summer. Each flower is constricted at the upper part of the tube and flares into scalloped lobes at the base, creating a vase shape of intricate beauty. This is a fine plant for light shade, where it is tolerant of summer watering; it will also take full sun.

Dichelostemma volubile. (FRAN COX)

Triteleias for the Garden

Triteleias grow from flattened corms covered with straw-colored, patterned corm coats. The mature leaves have a distinct keel on the undersurface. The large, showy flowers lack the waxy appearance of those of the genus *Brodiaea*, but have a conspicuous midvein on each petal. Unlike true brodiaeas and dichelostemmas, some triteleias have yellow or white flowers.

Some triteleias are easy to grow, while others are slow to multiply and establish in the garden. If their cultural requirements are met, many will naturalize by self-sowing, forming drifts among grasses and shrubs and in open woodland. Some are tolerant of summer watering. Triteleias are effective in pots as well, where they can be prominently displayed when in flower and moved out of the limelight during the dormant season.

Triteleia bridgesii, Bridges' brodiaea, Bridges' triteleia
Prefers: low elevation, part sun, dry summer dormancy, good drainage
Blooms: late spring, early summer
Bridges' brodiaea has pink to rich blue-purple flowers with brilliant reddish blue anthers. The corolla throats have a reflective, shiny surface of translucent cells, giving them a glistening appearance. The flowers are in large, nicely proportioned heads about six inches across and open sequentially, providing a lovely display for several weeks. The plant is robust and easy to grow in gardens, where it prefers filtered sun and good drainage.

Bridges' brodiaea is native to the edges of mixed evergreen and coniferous forests and foothill woodlands, often on rocky soils, in the Klamath, North Coast, and Cascade Ranges and the Sierra Nevada foothills.

Triteleia hyacinthina, white brodiaea
Prefers: low to mid-elevation, sun, dry summer dormancy, good drainage
Accepts: some summer water
Blooms: spring
White brodiaea is a prolific grower, sometimes covering acres of low hills and valleys with its dense clusters of showy white flowers in spring. It adapts well to garden conditions. Variable in the wild, some plants have stout, broad umbels of thick-petaled, milky white flowers on tall scapes, while others bear delicate, dwarf,

Triteleia hyacinthina. (STEVE EDWARDS)

crystalline white flowers in smaller umbels on short scapes. The taller forms, usually over two feet tall with flower clusters four inches across, often have chalk-white flowers visible from some distance. Others have more compact umbels two to three inches in diameter. Rare blue-flowered forms have been found.

White brodiaea is native to coniferous forests, foothill woodlands, and valley grasslands in the Central Valley, Sierra Nevada, and Cascade Ranges to British Columbia and Idaho, usually on sites that are moist in spring.

Triteleia ixioides, golden brodiaea, pretty face
Prefers: low to high elevation, part shade, moisture, good drainage
Blooms: spring, early summer
Golden brodiaea is native to edges of coniferous forests, foothill woodlands, and valley grasslands in the Klamath and Cascade Ranges, the Sierra Nevada, and other parts of central western California into southwestern Oregon. It is elegant and airy, with umbels of starlike, light yellow flowers with contrasting bronzy to light

brown stripes on each narrow petal. A vigorous grower, it makes a pleasing contrast with the predominating blues and purples of other brodiaeas.

In montane areas, ***Triteleia ixioides* ssp. *anilina***, mountain pretty face, may flower in summer on short scapes from half an inch to two inches tall. This subspecies is more difficult to grow than the species, but may succeed with excellent drainage in shade with irrigation; it is effective in rock gardens or in containers. At lower elevations ***T. ixioides* ssp. *scabra***, foothill triteleia, is found in

Triteleia ixioides (SAXON HOLT)

scrub and grassland and tolerates more sun and dryness than the species. It flowers in spring on eighteen-inch to three-foot scapes.

Triteleia laxa, Ithuriel's spear
Prefers: low to mid-elevation, sun, dry summer dormancy, good drainage
Accepts: light shade, some summer water, clay soils

Triteleia ixioides

Triteleia laxa. (ROBERT CASE)

Blooms: spring Ithuriel's spear is probably the best-known triteleia and the easiest to grow in gardens, where it multiplies rapidly, even in clay soils. Native to open forest, woodland, and grassland on clay soils in central and northern California to southwestern Oregon, it is breathtakingly beautiful over acres of open or lightly shaded slopes in spring.

From the broad, grasslike leaves emerge stout scapes a foot to just under two feet high with pale lavender-blue, blue-violet, to dark blue flowers in loose spheres four to eight inches across. Each flower is one to three inches long, flaring gracefully into an open funnel. The petals have a glistening, crepelike surface. The anthers are powder blue. This species is variable, and some plants have rose-violet, deep blue-purple, or pure white flowers.

Triteleia laxa **'Queen Fabiola'** is the most readily available cultivar, and its light blue to corn-flower-blue flowers are sold commercially by florists. It blooms in late spring or early summer. Best in full sun for at least half the day, it is ideal for borders, rock gardens, and as a groundcover.

Triteleia laxa

Triteleia lilacina, foothill
triteleia

Prefers: low elevation, sun, dry summer dormancy, good drainage

Blooms: spring, late spring

Foothill triteleia is found on volcanic hills and mesas in the foothills of the Cascade Ranges, the Sierra Nevada, the Sacramento Valley, and the northeastern San Joaquin Valley. The soft-lilac flowers, with powder-blue stamens, have an astonishing jewel-like quality. In sun and with good drainage, this triteleia adapts well to garden cultivation.

Triteleia peduncularis, seep brodiaea, long-ray brodiaea

Prefers: low to mid-elevation, sun, vernally moist conditions, dry summer dormancy

Blooms: late spring, summer

As its common name suggests, seep brodiaea is from wet seeps and streamsides, so it appreciates occasional water in early summer if rains have ended by mid-spring. By mid-summer it should be kept dry. Blooming in late spring or summer, its eighteen- to forty-inch scapes are topped with funnel-shaped or spherical umbels of narrow, flaring flowers on long, thin pedicels. Petals are white on the upper surface and often striped or tinted blue-purple underneath, giving the plant a different appearance when viewed from above than when seen from the side.

Seep brodiaea multiplies freely where its cultural needs are met. It is native to damp grasslands, vernal pools, and streams, often on serpentine soils, in central and northwestern California.

Triteleia peduncularis

Bloomerias for the Garden

Bloomerias, or goldenstars, are similar to triteleias in appearance. They are taxonomically differentiated primarily by the filaments, which have cuplike appendages at the base. Common in heavy soils of valley grasslands and hillsides in central to southern California, they have several long, narrow leaves and an umbel-like inflorescence with ten to more than thirty flowers, rich yellow with a hint of golden orange, opening in late spring.

Bloomerias grow from corms that require summer dryness. Corm replication is slow, but propagation from seed is easy.

Bloomeria crocea, common goldenstars

Prefers: low to mid-elevation, sun, dry summer dormancy

Blooms: late spring

Common goldenstars is similar to *Triteleia ixioides*, but its flowers are displayed in more spherical umbels and the pedicels are almost thread-like, giving the plant an even airier effect. It is native to chaparral, grasslands, and open woodlands in central and southern California to northern Baja California. As with most brodiaeas, common goldenstars is effective naturalized in grassland and meadow plantings. Clustered in shrub borders in the dry garden, it is striking against gray-leaved plants such as saffron buckwheat (*Eriogonum crocatum*) or California poppy (*Eschscholzia* spp).

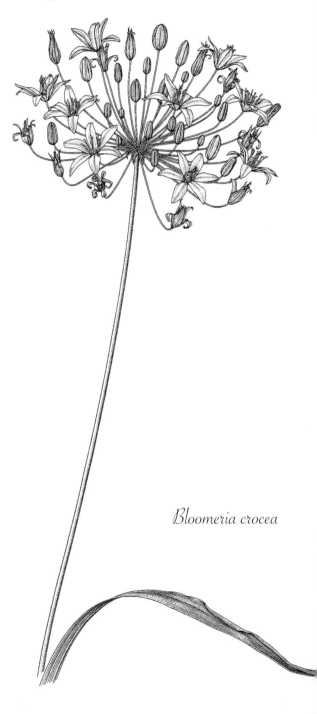

Bloomeria crocea

IRISES AND SISYRINCHIUMS

The iris family, Iridaceae, consists of some 1,800 species in about ninety genera from temperate and tropical regions of the world. The family includes about 300 species of the genus *Iris*, as well as many other plants familiar to gardeners such as crocus, freesia, gladiolus, and watsonia. Irises native to California, commonly known as Pacific Coast irises, easily hold their own in this showy company. They share the traits of other plants in the genus, including grasslike foliage, upright flower stems, and flowers with three petals and three sepals in bright colors, often with intricate markings.

Pacific Coast irises are all "beardless" irises. Their closest relatives probably are the Siberian irises (Sibericae series), which are native to Japan and China. Beardless irises are subdivided into at least eight series. Almost all of those native to California are members of the Californicae series. Two, *Iris missouriensis*, which grows in the Sierra Nevada, and *I. longipetala*, found in lowland western California, are in the Longipetalae series. *Iris tenuissima*, in the Evansia series, is native to northwestern California and southwestern Oregon.

Sisyrinchiums are diminutive members of the iris family with clumps of narrow, strap leaves and an abundance of open, six-sided flowers. They are widely distributed in grassy meadows, mostly near the coast. Some species require ample moisture; others tolerate some dryness.

IRISES

No members of the iris family are more beautiful than the Pacific Coast native irises. Their arching, narrow leaves and slender flower parts give them a delicate, sprightly woodland grace. Some flowers may appear in late winter, but blooming typically occurs in early spring to late spring. Each flower lasts only a few days, but another soon replaces it during the three- to four-week flowering season. The foliage usually ranges from ankle height to just below knee height.

Growing Irises

Cultural preferences and tolerances of native Pacific Coast irises vary, but most thrive in dappled shade or morning sun. Many will take full sun along the coast, but in hot inland gardens they should be planted in light shade. Deep shade encourages vegetative growth but discourages flowering.

Pacific Coast irises grow best where summers are long and dry, but they need winter rain and tolerate light snow cover and some frost. Soils should be neutral to slightly acid, lightly improved with organic matter, and watered infrequently. In the wild, soils where native irises grow tend to be on the acid side and about forty percent gritty material (the size retained by an eighth-inch screen).

Many Pacific Coast irises will grow in serpentine soil. Fertilizer is unnecessary, but can be applied lightly to established plants in fall. All native irises are reputed to dislike animal manure or lime. All are virtually ignored by deer, although gophers and rabbits may nibble them. Iris borers can decimate a patch, but seem to do conspicuous damage only every few years, giving time for plantings to recover. Native irises do not require periodic dividing as do bearded irises, although they can be propagated by division when new growth begins in fall.

If the soil is sticky clay, it is worth the effort to create an amended planting bed of well-drained, even rocky soil that sheds water above the adjacent grade. Soil sulfur may be used to reduce pH if it is 7.0 or higher. Rhizomes should be planted shallowly to discourage rotting.

Most Pacific Coast irises need a dormant period in summer, when watering should be greatly reduced or stopped, since they are vulnerable to fungus diseases from a combination of heat and moisture. This makes them difficult garden subjects in areas with summer rainfall. In the Coast Ranges, the best results with the Californicae (see p. 96) have been obtained by withholding water entirely in summer after plants are established. In hotter climates, plants do best with occasional deep watering, with the soil near the rhizome kept as dry as possible. Browning leaves during summer dormancy may be trimmed back or left on the plant.

Pacific Coast irises grow easily from seed. Viability of seed declines in the first few years, but continues for up to ten years. Seeds should be sown about one-quarter inch deep in a fast-draining potting mix and kept moist and cool until germination, which

Iris
douglasiana
Seed pods

takes about two months. Seedlings can be moved to the garden
or into individual pots when they
are three to six inches tall. They
should be watered sparingly during the first summer, and usually
will bloom the second or third year
from seed. Flower quality may not
be true in the first blooming, so
undesirable seedlings should be
culled only after their second
bloom.

Iris
douglasiana
Division

Rhizomes of adult plants can
be divided if clumps become
crowded or new plants are desired. This can be done by lifting
the whole clump and gently break-
ing it apart by hand at natural

Iris in garden setting. (SAXON HOLT)

separations. Divisions should be replanted at the same level as the plant from which they were taken. Plants should be divided only when new white roots are starting to form or are in active growth, not when plants are dormant. Without active root growth, the divisions may not survive. Growth of new roots can be stimulated by watering ahead of fall rains.

Pacific Coast irises are effective planted with other perennials, as specimens, or massed as tall groundcover in dry or semi-dry shade of acid-loving trees and shrubs. They also provide vertical accents among low groundcovers. Native irises are ideal plants for the water-conserving garden. The grassy foliage clumps look good in combination with other Mediterranean-climate plants that thrive in dry, dappled shade. Many Pacific Coast irises grow well in the dry, rooty shade of mature shrubs and trees. They combine effectively with bunchgrasses, hardy geraniums, dicentras, violas, and groundcovers such as a mix of low-growing sedums.

The flowering season of Pacific Coast irises can be prolonged by planting a combination of early, mid-season, and late-blooming species and varieties. All can be grown in well-drained soil in porous containers, although they will require occasional division and more watering than if planted in the ground. Many native irises are excellent for erosion control on steep slopes, as they can form large mats to hold loose soil. Some of the smaller irises, such as *Iris innominata* and some plants of *I. macrosiphon*, are good rock garden plants.

Rot and rust diseases sometimes occur when California irises are grown away from their natural habitats. Natural soils contain an array of beneficial and harmful microorganisms. The beneficial microorganisms are dominant in healthy soil. A microbial inoculant of beneficial bacteria and fungi has improved growth in problem locations at the Santa Barbara Botanic Garden, presumably by keeping harmful and beneficial organisms in balance.

In addition to the species, there are many named hybrids in a wide range of colors that offer exciting possibilities for the garden. Because many native species hybridize readily, care should be taken to plant hybrids only in settings where their pollen will not pollute a local wild population.

Irises for the Garden

Eleven Pacific Coast iris species are such a distinctive group that they form the Californicae series within the beardless section of the irises. Most plants in this series are found in California, with a few from farther north. Their flowers have the three upright standards and three spreading falls common to the genus. However, colored veins rather than beards function as nectar guides.

In terms of cultural preferences, Pacific Coast irises may be divided into three groups: those that need regular summer watering (*Iris missouriensis*); those that prefer infrequent summer watering (*I. bracteata*, *I. chrysophylla*, *I. douglasiana*, *I. innominata*, *I. longipetala*, and *I. tenax*); and those that, once established, can go without water all summer (*I. fernaldii*, *I. hartwegii*, *I. macrosiphon*, *I. munzii*, *I. purdyi*, and *I. tenuissima*). These characteristics, combined with other features, determine their overall suitability for garden use.

Pacific Coast irises are widely cultivated in gardens in the western states, and a few varieties and many cultivars are commonly available in nurseries. Seeds and plants of species and cultivars also are available from plant societies and seed exchanges.

Iris bracteata, Siskiyou iris
Prefers: mid-elevation, part shade, moisture, good drainage
Blooms: spring
A showy plant, Siskiyou iris has large, creamy white to golden yellow flowers with conspicuous reddish brown veins. Two flowers usually are borne on each nine- to twelve-inch stem. The deep green leaves are broad, thick, and stiff, up to two feet tall and one-half inch wide. A sparse, slow spreader, this iris is partly deciduous in winter. It is native to shady or sunny spots in grassy flats and pine forests of the Siskiyou Mountains in Del Norte County and southwestern Oregon, often on serpentine substrate.

Siskiyou iris is on the CNPS List 3 of potentially endangered plants for which more information is needed. Seeds are occasionally available from specialty nurseries and seed exchanges.

Iris chrysophylla, yellow-flowered iris, slender-tubed iris
Prefers: mid-elevation, part shade, moisture, good drainage
Blooms: spring
Yellow-flowered iris is native to road banks, meadows, and exposed, sunny sites in open coniferous forests in the western and southern Klamath Ranges in northwestern California and western Oregon.

This diminutive plant has delicate, flattish flowers of creamy white or off white with gold, reddish brown, or violet veining, usually two on each five-inch stem. Flower parts are narrow, and the style crests are long and slender. Slender, light green leaves are dull-surfaced and six to eight inches long. Flowers can be abundant and last two days each. In the wild, yellow-flowered iris usually grows in the light shade of conifers. In the garden it needs occasional summer water.

'**Valley Banner**' is a mid-season hybrid of *Iris chrysophylla* and *I. tenax* collected in the wild by Ruth Hardy in 1958. Its elegant flowers are white with purple veins and styles.

Iris douglasiana. (STEPHEN INGRAM)

Iris douglasiana, Douglas iris
Prefers: low to mid-elevation, cool, coastal habitats, sun to part shade, dry summer dormancy
Accepts: heat, some summer water inland, clay soil
Blooms: late spring

This is probably the best all-round Pacific Coast native iris for cultivated gardens, especially as a groundcover, and it is the most commonly available in nurseries. Douglas iris grows on bluffs and open grassy hillsides in coastal prairie and in mixed evergreen forest along the Pacific Coast from Santa Barbara to central Oregon. It is unpalatable to livestock, and although some ranchers are said to consider it an aggressive weed, agricultural commissioners in north-coastal counties report that this species is not a significant problem in rangeland. Douglas iris hybridizes readily with other Pacific Coast irises, and some natural hybrid populations have been given their own names, such as the "Marin iris" of the Coast Ranges north of San Francisco.

Douglas iris is generous with its flowers, each branched stem bearing two or three blossoms. Flowers are light blue-violet to dark purple, sometimes creamy white or pure white, rarely light yellow. Plants form robust clumps of handsome, evergreen, dark green leaves, glossy on the upper surface and dull green below, usually with a reddish base.

Iris douglasiana

Easily grown, readily transplanted, and quite adaptable to garden conditions, this iris tolerates the combination of heat and some summer moisture that is fatal to many other native irises. Outside the coastal fog belt it requires some summer irrigation, and in hot interior locations it should be given some shade.

Douglas iris is widely used as parent stock for hybrids, and there are many garden-worthy selections.

Iris douglasiana 'Canyon Snow'.
(CAROL BORNSTEIN)

'**Canyon Snow**' is an early to mid-season, vigorous seedling selected by Dara Emery at the Santa Barbara Botanic Garden in 1974. The flowers are pure white with yellow streaks or spots on the falls, and its sturdy, upright, exceptionally shiny, rich green foliage is highly rust-resistant and virtually evergreen.

'**Agnes James**' is an early-blooming, white-flowered selection collected in the wild in southwestern Oregon by Carl Starker and registered in 1939.

'**Amiguita**' is an early to mid-season bloomer selected in 1947 by Eric Nies. Its flowers are ruffled and blue bi-tone with a dark purple signal spot on the falls.

'**Harland Hand**', with purple flowers, has a long blooming season.

'**Mendocino Banner**' has white flowers with purple veins and contrasting purple style crests.

Iris fernaldii, Fernald's iris

Prefers: low to mid-elevation, part shade, dry summer dormancy, good drainage
Blooms: late spring, early summer
This iris bears buff to pale yellow, creamy white, or pure white flowers with delicate gold and lavender veining. Flowers are held upright on sturdy stalks. Foliage is blue-green or gray-green and eight to sixteen inches long. Plants tend to travel rather than clump.

Native to the Central and North Coast Ranges in rich, humusy soil in mostly shaded woodland and on grassy slopes, the species crosses readily with other native irises and is being replaced in some locations by spontaneous hybrids. Some of these hybrids tolerate more sun than the species.

Iris hartwegii, Sierra iris

Prefers: mid- to high elevation, part shade, dry summer dormancy, good drainage
Blooms: late spring, early summer
Sierra iris is a two- to sixteen-inch-tall, deciduous iris that usually displays one or two light cream to bright yellow or sometimes lavender flowers per stalk in late

spring. It is fairly common in sunny, open or partly shaded sites in mixed woodland at mid- to high elevations above the foothills of the Sierra Nevada between Kern and Plumas Counties. It is difficult to maintain in lowland gardens.

There are four distinctive subspecies of Sierra iris. *Iris hartwegii* **ssp.** ***pinetorum*** bears creamy white or yellow flowers with gold veining. *Iris hartwegii* **ssp.** ***australis***, a diminutive high-elevation native of the San Bernardino Mountains with lavender or purple flowers, is rarely grown by gardeners. *Iris hartwegii* **ssp.** ***hartwegii*** has pale to golden yellow or lavender flowers and grows in the southern Cascade Ranges and the Sierra Nevada. The cold-hardy subspecies *I. hartwegii* **ssp.** ***columbiana***, Tuolumne iris, with pale yellow flowers, is on the CNPS List 1B of plants considered rare, threatened, and endangered in California and elsewhere. Seeds are occasionally available from plant society seed exchanges and specialty seed suppliers.

Iris innominata, Del Norte County iris

Prefers: mid-elevation, inland, part shade, moisture, excellent drainage
Accepts: regular summer water in cool locations
Blooms: early spring

Iris innominata. (SAXON HOLT)

Iris longipetala

This woodland iris forms compact, almost miniature tussocks of dense, evergreen, cold-hardy foliage to eight inches tall. Its small stature makes it a choice rock garden plant. Seldom found along the coast, it is locally abundant inland on serpentine soils in southern Oregon and in Del Norte County in adjacent northern California. Because of its limited distribution, it is on the CNPS "watch" list (List 4) of potentially endangered plants. Seeds and plants are available from specialty nurseries and seed suppliers.

Because of its clump-forming habit, compact evergreen leaves, and abundant flowers, *Iris innominata* is often used for hybridizing, and many plants sold under this name in nurseries actually are hybrids. Leaves are fine and grasslike, dark glossy green above and lighter below. The numerous flowers usually are bright golden yellow with darker veins, but they vary from creamy white to deep gold and from orchid pink to deep blue-purple, with brown, red, or purple veining. Unfortunately, the luscious yellows fade readily. Flower stems, usually about five inches tall, each bear a single flower (or sometimes two).

Although long known to residents of its native area, *Iris innominata* was not officially named until about 1930 (*innominata* means unnamed). Where summers are not too hot, this iris can take occasional to regular summer watering. Good drainage is a must for this species and for hybrids in which it is a parent.

Iris longipetala, long-petaled iris
Prefers: low elevation, sun, coastal habitats, moist, good drainage
Accepts: summer dryness along coast
Blooms: spring
This is essentially a low-elevation form of *Iris missouriensis*, and it is in the Longipetalae rather than the Californicae series. Best with summer fog or intermittent watering, it will survive summer dryness except in hot inland areas. The flowers are strongly veined in purple on a lighter background, and the one- to two-foot-long, gray-green leaves are briefly deciduous. Long-petaled iris is found in moist, grassy meadows along the north-central coast and in the San Francisco Bay Area. It is sometimes a conspicuous feature of grazed pastures, because its foliage is unpalatable to cattle. It is quite adaptable to garden conditions, although probably not hardy below 10 degrees F.

Iris macrosiphon, ground iris, long-tubed iris
Prefers: low to mid-elevation, full or nearly full sun, dry summer dormancy
Blooms: spring

Ground iris is widespread in sunny grasslands, meadows, or open woodlands in the Coast Ranges and the Sierra Nevada foothills from central California south to Santa Cruz County. Widely variable, it generally has dull-surfaced, narrow, blue-green leaves and lightly fragrant flowers that range from deep blue, lavender, or purple to creamy white, pure white, or occasionally deep golden yellow.

Over time, as seeds from several generations fall in the same area, plants form good clumps up to ten inches tall. Plants in full sun often have very short stems. The flowers, two to a stem, nestle low in the foliage or are carried slightly above it.

Iris macrosiphon resembles a lighter-leaved *I. innominata* and can be substituted for that species where summer water is not available.

The cultivar **'Mt. Madonna'**, introduced by Suncrest Nurseries in Watsonville, has violet flowers.

Iris missouriensis, western blue flag
Prefers: mid- to high elevation, sun, wet places
Blooms: early summer
Western blue flag, in the section Longipetalae, is the one true wet grower among the California native iris species. It inhabits moist meadows, and in the garden it needs regular summer irrigation, at least until flowering. After flowering, intervals between waterings should be lengthened to allow partial dormancy. A deciduous hardy iris, its light green leaves, eight to twenty inches

Iris macrosiphon

long, are more upright than those of most other California irises. Flowers are pale lilac to almost white with lilac-purple veins.

Found throughout much of the West at mid- to high elevations on gentle slopes that are moist most of the year, this is an easy and satisfactory garden plant for higher elevations. Along the coast and at low elevations, plants do not persist and rarely flower. In pastures that are overgrazed for a long period, it may become well established because its leaves are unpalatable to livestock.

Iris munzii, Munz's iris
Prefers: mid-elevation, part shade, cool, moisture, good drainage
Blooms: spring
Munz's iris bears large, classically proportioned pale lavender, purple, or reddish purple flowers, two to four to a stem. Leaves are blue-green, rigidly upright, wide, and up to two feet long. Plants do not form spreading clumps as do many other irises, but grow slowly as stately individual plants. *Iris munzii* is not reliably hardy below 15 degrees F., and it is especially prone to rust in cool, humid climates. Rust can be controlled by removing old leaves just before new foliage appears in fall.

Native and endemic to woodlands and moist, moderately shaded streamsides in the Sierra Nevada foothills of Tulare County, Munz's iris can take some summer irrigation as long as it is allowed to dry out between waterings. It is on the CNPS List 1B of plants rare, threatened, or endangered in California and elsewhere. Seed-grown plants and seeds occasionally are available from plant societies and seed exchanges.

Some plants derived from this species have the clearest blue flowers of the Pacific Coast native irises, with a pale and glowing, sometimes almost turquoise quality known as "Munzii blue." This trait originated in a series of selections and hybrids made by Lee Lenz at Rancho Santa Ana Botanic Garden in the early 1970s, most of which were lost during subsequent changes at the garden.

'Sierra Sapphire', an early to mid-season pure *Iris munzii* seedling selected by Lee Lenz in 1972, displays this distinctive true-blue quality, but plants with such flowers have not been found again in the wild.

Iris purdyi, Purdy's iris
Prefers: low to mid-elevation, part shade, moisture, good drainage
Blooms: spring
Purdy's iris has large, flattish flowers, almost resembling a hybrid clematis, that distinguish it from other native irises. The white, creamy white, or soft yellow

petals have prominent reddish brown, cerise, or purple veins. Purdy's iris is native to redwood forest and mixed evergreen forest in filtered sunlight and humusy soil in the North Coast and Klamath Ranges from Sonoma to Trinity Counties.

Mature plants grow singly or in loose, sparse clumps of glossy, deep green leaves. In much of its natural range, Purdy's iris has crossed with other iris species, and the resulting hybrids are quite variable.

Iris tenax ssp. *klamathensis*, Orleans iris, Klamath iris

Prefers: mid-elevation, sun or light shade, some summer water, excellent drainage

Blooms: early spring

Iris tenax, Oregon iris, was introduced into England in 1825 by explorer-naturalist David Douglas, and was the first Pacific Coast iris in cultivation. Its narrow, light green, evergreen leaves, to sixteen inches long, grow taller than the flower stalks. Stalks usually bear single flowers in a range of colors, from purple or lavender to white, creamy white, or yellow. The most northerly and cold-hardy Pacific Coast iris, *I. tenax* is native to Oregon and Washington, where it is found in open or lightly shaded sites in woodlands or sunny streamside clearings. It may not prosper in more southerly gardens.

The California subspecies *Iris tenax* ssp. *klamathensis*, Orleans iris or Klamath iris, occurs in the western Klamath Ranges near Orleans in Humboldt County. It is of limited distribution and is on the CNPS "watch" list (List 4) of potentially endangered plants. This plant prefers more shade than the species and is used by breeders to improve the cold-hardiness of hybrids. Seeds occasionally are available from botanic garden and plant society sales or seed exchanges.

Iris tenuissima, long-tube iris, slender iris

Prefers: low to mid-elevation, shade, dry summer dormancy, good drainage

Blooms: late spring, early summer

Long-tube iris is from shaded, duff-covered forest floors or dry, sunny openings in mixed evergreen forests and oak woodlands of northern California foothills around the upper end of the Central Valley and the Cascade and southern Klamath Ranges. Adaptable to garden cultivation, this iris bears creamy white to straw-colored, pale yellow, or white flowers, two to each stem, with falls veined purple or brown and long, reflexed style crests. Flower parts are slender, and petals and sepals spread outward, their edges often ruffled. The foliage is usually less than sixteen inches long.

Pacific Coast Iris Cultivars

Where the habitats of Pacific Coast iris species meet, either naturally or as a result of human disturbance, spontaneous hybrids often occur. Naturally occurring hybrids have been reported for all the Californicae series except *Iris munzii*.

The high degree of interfertility among species also has provided opportunities for hybridizers. Their work has increased length of blooming season or extravagance of display; disease resistance; climatic tolerance; tolerance of watering in cultivation; and ruffles, veining or patterning, and color range. Nearly 1,000 hybrid Pacific Coast irises have been introduced over the past few decades, and new variet-

Iris hybrid. (SAXON HOLT)

ies are judged each year at local, regional, and national shows. A few notable selections and hybrids include the following.

'Chimes' and **'Fairy Chimes'** are both early to mid-season hybrids of unknown parentage registered by Jack McCaskill of Pasadena in 1972. Flowers are ruffled and creamy white with gold markings. Both reportedly thrive in difficult situations. 'Fairy Chimes' is especially floriferous.

'Claremont Indian' is an early-blooming, light mahogany-red-flowered progeny from seedlings raised by Lee Lenz, registered in 1955, and a source for later red-flowered plants.

'Garden Delight' is a late-blooming, light yellow-flowered plant with a large medium-brown blaze on the falls. Its parentage includes 'Western Queen', 'Ojai', and 'Claremont Indian'. It was registered by George Stambach in 1971.

'Iota' is a cross between *Iris purdyi* and *I. tenax* made by William Dykes of

England. This was the first California iris given an Award of Merit by the Royal Horticultural Society (1914).

'**Native Born**' is a late-blooming iris derived from a hybrid line by Joe Ghio of Santa Cruz in 1971. Its flowers are purple with a cream-colored thumbprint.

'**Native Warrior**' is an early-blooming near-red hybrid of 'Amiguita' and 'Claremont Indian' registered by August Phillips in 1970.

'**Ojai**' is an offspring of 'Amiguita' registered in 1959 by Marion Walker of Ojai and widely used by other hybridizers. Its flowers are light purple-violet with a darker blaze around the yellow signal.

'**Pacific Moon**' is an early-blooming hybrid registered by Ben Hager of Stockton in 1973. Its flowers are creamy white with lavender veins. This plant is often used in breeding.

'**Restless Native**' is an early to mid-season hybrid with flowers of two-tone red registered by Joe Ghio in 1970.

'**Soquel Cove**' is an early bloomer from hybrid lines and collected blue-flowered plants of *Iris munzii* registered by Joe Ghio in 1976. Its white flowers have a turquoise wash.

SISYRINCHIUMS

Sisyrinchiums are diminutive plants with irislike foliage and flat-faced flowers that are displayed in profusion on sunny days in spring, closing in dark or cloudy weather. Flowers consist of six pointed, upward-facing tepals and are about half an inch across. Generally short-lived and somewhat wispy in the wild, sisyrinchiums can become robust clumpers in the garden. All need full or nearly full sun. They are effective in small-scale gardens, rock gardens, or in containers. They self-sow, sometimes freely, and also can be propagated by division of rhizomes.

Sisyrinchiums for the Garden

Sisyrinchium bellum, blue-eyed grass
Prefers: low to mid-elevation, sun, some moisture, good drainage
Accepts: summer dryness, part shade, winter cold, clay soils, seasonal flooding
Blooms: spring, early summer
Blue-eyed grass is native to open, seasonally moist woodlands and coastal to interior grasslands throughout much of California and into Oregon. It tolerates diverse soils, from sandy to clay.

Sisyrinchium bellum

Sisyrinchium bellum. (SAXON HOLT)

The narrow, grassy leaves are eight to twelve inches tall, with flower stems somewhat taller. Flowers are pale blue to purple with yellow centers. In the garden, plants can go dry in summer, but also accept summer watering. Selections include dwarfs only a few inches high, with flower colors ranging from white to pale blue or sapphire blue to lavender or deep purple. The jointed flower stems of blue-eyed grass can be rooted like cuttings.

Plants and seeds are widely available in specialty nurseries, at plant society sales, and from mail-order seed exchanges. There are a number of named cultivars offered, including **'Greyhound Rock'**, with pale blue flowers and gray-green foliage; **'Rocky Point'**, with short, broad leaves and purple flowers; **'E. K. Balls'**, similar to 'Rocky Point' but with taller flower stems; **'Album'**, with white flowers; and **'Nanum'**, a dwarf variety. **'California Skies'** has light blue to bright blue flowers and grayish green leaves. **'Figueroa'** has a creamy variegation on the leaf margins and blue flowers.

Sisyrinchium californicum, golden-eyed grass, yellow-eyed grass
Prefers: low elevation, coastal habitats, sun, year-round moisture
Accepts: seasonal flooding
Blooms: late spring to late summer
Golden-eyed grass has irislike leaves and bright yellow flowers. Native to seeps and other wet places in the San Francisco Bay Area and north coastal California to British Columbia, it is best with year-round moisture and full sun and can grow in wet or poorly drained spots in the garden. Easy to grow, it is effective along the edge of a planting bed or in containers. It is not tolerant of drought or much cold,

but where conditions are right it self-sows freely and can be invasive. Spent foliage turns black but is easily removed.

Sisyrinchium douglasii var. *douglasii* (*Olsynium douglasii*), purple-eyed grass
Prefers: mid- to high elevation, sun to part shade, moisture, good drainage
Blooms: spring
Purple-eyed grass is native to open, often rocky, vernally moist slopes in coniferous forests of the Klamath, North Coast, and Cascade Ranges to British Columbia, Idaho, and Utah. Slow to multiply and not as easy in the garden as other sisyrinchiums, it has small tufts of narrow leaves and reddish purple flowers.

Sisyrinchium elmeri, Elmer's golden-eyed grass
Prefers: mid- to high elevation, sun, moisture, good drainage
Blooms: spring
This plant is native to moist meadows and coniferous forests in the southern Klamath and Cascade Ranges and the Sierra Nevada to the San Bernardino Mountains in southern California. It resembles *Sisyrinchium californicum* but is more compact and has smaller, less showy, deep yellow or orange-yellow flowers. It is also more difficult to grow at low elevations. Best in moist locations, it blooms repeatedly in hot weather. The yellow flowers have a brown spot at the base and give a lacy effect in mass.

Sisyrinchium idahoense, Idaho blue-eyed grass
Prefers: mid- to high elevation, sun, year-round moisture, excellent drainage
Blooms: late spring, summer
Native to moist, high mountain meadows in the Klamath and Cascade Ranges, the Sierra Nevada, and northeastern California to British Columbia, Montana, and Colorado, *Sisyrinchium idahoense* prefers year-round moisture and requires excellent drainage. It makes a fine rock garden plant. Flowers are blue-violet, sometimes white, and the grayish leaves are broader than those of other sisyrinchiums. Plants self-sow. Seeds and plants are often available from alpine plant societies and seed exchanges. The cultivar **'Album'** has white flowers.

ARUMS

The arum family, Araceae, is a large family of mostly tropical and subtropical, terrestrial or aquatic perennials that grow from rhizomes, corms, or tubers. The family abounds in strange-looking oddities and is well known through such warm-climate plants as philodendron, calla, caladium, anthurium, dieffenbachia, and jack-in-the-pulpit.

Several members of the family have naturalized in some parts of California and the Northwest, including sweet flag (*Acorus calamus*), arum (*Arum italicum*, *Peltandra virginica*), water-lettuce (*Pistia stratiotes*), and calla lily (*Zantedeschia aethiopica*). Dozens of others are commonly cultivated, mostly as indoor or greenhouse plants. One temperate species, skunk cabbage (*Lysichiton americanus*), is native to north coastal California, including the San Francisco Bay Area.

Lysichiton americanus, yellow skunk cabbage

Prefers: low to mid-elevation, coastal habitats, humus-rich soil, wet places, shade to part sun

Blooms: spring

Lysichiton americanus. (ROBERT CASE)

The genus *Lysichiton* consists of two marginal aquatic species, one in northeast Asia and the other, *L. americanus*, which ranges from north coastal California to Alaska, Montana, and Idaho. At home at the margins of slow-moving or stagnant water, yellow skunk cabbage is a bold and dramatic plant. A stem (spadix) bearing tiny green flowers is surrounded by a daffodil-yellow sheath (spathe) up to eighteen inches tall that usually precedes the emerging leaves. The flowers have a heavy, musky scent that some find unpleasant. Forming a basal rosette up to three feet tall and wide, the large, broad, rich green leaves with a thick midvein have a leathery texture and a slight sheen, providing a lovely backdrop for other moisture-loving plants.

In California *Lysichiton americanus* inhabits boggy areas, streamsides, and river banks in cool forests near the coast, usually in deep shade but sometimes in part to full sun. Effective in drifts around pools or along streams, plants seed freely and in time can make a lush and rewarding display in the spring and summer garden. Plants die back in autumn. Seed can be collected and sown, but propagation by division of the thick, fleshy rhizomes is easier. Divisions take a couple of years to bloom.

ORCHIDS

The orchid family, or Orchidaceae, is a vast group of plants, currently estimated at over 800 genera and 18,000 species. Of these, the overwhelming majority are native to the tropics, and the temperate species are distributed primarily in regions with summer rainfall. California hosts only a few as dazzling as the fringed orchids (*Platanthera* spp.) and grass pinks (*Calopogon* spp.) found along the East Coast, or the many subtropical orchids that grow in southern Florida. However, these few— at least those that can be cultivated at all—are well worth the gardener's efforts. Flowers of most California species are small but interestingly formed and colored, inviting close inspection.

As might be expected from their sheer numbers, orchids worldwide are an extremely variable group. The stems are often bulbous, forming durable storage organs, although many of the temperate orchids have simple, seasonal stems. The leaves are always undivided, but they can take a variety of forms. Most terrestrial orchids have palmlike, lilylike, or grassy leaves.

Orchid flowers are unique in form. The three sepals are often regular, as are two of the three petals, but the third is modified, often into fantastic structures collectively called the lip or labellum. In place of the typical stamens and pistil of other monocots is an odd, often hooded or crested structure called the column. These features, and the often unusual coloring of the flower parts, make all orchids interesting, and some of them spectacular.

Nearly all orchids have tiny, almost dustlike seeds containing virtually no stored food. They are dependent on certain fungi for nutrients as they sprout and grow.

Growing Orchids

California's few native orchids are delicate both in appearance and in physical structure. Overgrowth by more vigorous plants such as carpeting perennials and shrubs or grasses is a constant danger. Human and animal traffic also poses a considerable threat. All of this suggests planting native orchids by themselves or with other delicately textured, well-behaved plants such as small bulbs and ferns. Mossy mounds, banks, and other elevated sites are ideal, as are spots in a well-tended rock garden, where plants are uncrowded and easily viewed.

Forest-dwelling orchids are beautiful in combination with wood fern (*Dryopteris arguta*), bleeding hearts (*Dicentra* spp.), heucheras, fairy lanterns (*Calochortus* spp.), and trilliums, all in the shade of an overhanging tree. Those that inhabit sunny streamsides and moist meadows can be interplanted with other denizens of similar sites, such as small mountain sedges and grasses, or placed by naturalistic ponds, always with care to maintain good soil drainage. Several native orchids are good subjects for containers, preferably unglazed clay pots, which provide good air circulation about the roots. These can be placed on elevated benches or suspended so that the plants may be admired at close range.

California native orchids occupy unusual habitats in the wild, and these must be closely replicated for their successful culture. Most grow in forest duff or on gravelly slopes and ledges where there is excellent drainage or even continual movement of water, plus good air circulation around the roots. These conditions inhibit the growth of predatory fungi, to which most orchids are susceptible. Most California orchids have either a year-round supply of moisture or summer shading of the roots, which keeps soil temperatures down and retards water loss from delicate tissues. Plants in the wild are shielded from snails, slugs, and rodents by rocks and other vegetation, and these pests also are kept in check by their own natural enemies (something lacking in the typical garden).

Some of these conditions can be imitated in the garden by mixing local soil with peat moss and a coarse sand, fine gravel, or perlite. One part soil, one to two parts peat, and two parts of the rocky component should yield a good mix, suitable even for species found naturally in heavier soils. Pot culture is a practical means of growing several native orchids, and it provides an easy means of protecting them from foraging animals. However, it requires even coarser mixes than one would use in the open ground to compensate for restricted drainage and air circulation. Whether in pots or in the ground, soil around the roots and crowns must be kept cool to inhibit the growth of water molds, which can kill whole plants in a few days' time. Where summers are constantly warm to hot, plants should be kept in the shade. Pots can be moved around as the seasons change to take advantage of any shady nook.

Most orchids require more or less continuous moisture. Fresh water low in dissolved salts is desirable, even essential for some species. None of the California native orchids requires heavy feeding; in fact, this can be quite detrimental, encouraging floppy growth and making plants more prone to disease. Only occasional applications of a balanced fertilizer, at about half the rate recommended for most plants, will suit them well; high-nitrogen fertilizers should be avoided.

Certain native orchids—including some of our showiest species—have failed repeatedly in cultivation for various reasons, only some of which are understood. Orchid fanciers have persisted over the years in collecting plants and attempting to solve their riddles. These attempts have merely hastened the decline of orchids in the wild, spectacularly so in the case of the lady's-slippers (*Cypripedium* spp.).

Propagation of most California native orchids is in its infancy, yet it may well be key to their preservation, as habitat destruction and collection take their toll. Only a few species—notably *Epipactis gigantea*—have the well-branched rhizomes needed for propagation by division. Because of their strongly seasonal cycles of both shoot and root growth and the delicacy of their stems and leaves, division should be attempted only in fall and winter, when plants are dormant. Rhizomes can be cut with a clean, sharp knife or shears to make pieces with multiple growth "eyes" or shoot buds. Roots should be teased apart with care to minimize breakage, and the pieces should be replanted immediately.

Epipactis gigantea
Seed Pods

The problems inherent in seeding orchids—the tiny seeds and the lack of stored food reserves—were solved long ago for tropical and subtropical orchids. Although current techniques involve gel and liquid media with sugars and other nutrients, clean facilities, and sterilized containers and tools, they are regularly practiced by private hobbyists. Seeds are available from some seed exchanges, but grown from seed, plants may take many years to flower. Seedlings, and even flowering-sized plants, occasionally are offered by mail order from commercial laboratories.

Tissue culture is another technique applying similar growth media and equipment to the propagation of tiny pieces extracted from the shoot tips of live plants. Tissue culture can yield almost unlimited quantities of particular clones, but California's native orchids have not yet been of sufficient commercial importance for this technique to be fully explored or widely utilized.

Orchids for the Garden

The following native orchids are candidates for garden cultivation if their exacting requirements can be met. They are not often available in the nursery trade, but some specialty growers occasionally offer them for sale.

Cypripedium, lady's-slipper
Prefers: low to high elevation, cool, moist, acid soil
Blooms: early to mid-summer
Lady's-slippers are widely distributed in the northern hemisphere but threatened almost everywhere by habitat destruction and collecting. They grow from tough, often branched rhizomes, thickly lined with long, fleshy roots that have an odd, somewhat musky odor. Basal leaves are oblong to lanceolate, often ribbed. Depending on the species, each seasonal shoot may have from a single pair to many large, deeply veined, nearly round to pointed-oval leaves. One to several unusually formed, often showy flowers are borne at the shoot tips from late spring to summer.

Two California *Cypripedium* species, *C. californicum* and *C. montanum*, both summer-blooming, deserve to be cultivated in gardens, although neither is easy to grow and both are only occasionally available from specialty suppliers. The beauty of these two orchids has tempted many gardeners to try to grow them, but few have succeeded for long, perhaps because of their susceptibility to fungus invasion. These cold-hardy plants need loose, acid soil, a cool root run, and constant moisture to thrive.

Cypripedium californicum, California lady's-slipper, inhabits sunny, gravelly seeps, bogs, and streambanks in the Klamath, North Coast, and Cascade Ranges and the northern Sierra Nevada to southwestern Oregon. Plants may grow to four feet tall, with bright green stems and equally bright green, two- to six-inch leaves. In the garden, clumps may form under the right conditions, which include semi-shade, humusy soil with good drainage, and abundant water until flowering.

From one to a dozen or more delightfully fragrant, greenish yellow flowers are borne on each stem. Individual flowers are usually a little less than an inch wide and modestly colored, with fuzzy tan to chartreuse sepals and petals and a white to pinkish white lip with brown spots. Uncommon in the wild, *C. californicum* is on the CNPS "watch" list (List 4). It is occasionally available from specialty suppliers.

Cypripedium montanum, mountain lady's-slipper, is an uncommon native of moist or dry mixed evergreen or coniferous forests and subalpine slopes in the

Cascade Ranges, the central and northern Sierra Nevada, and the Modoc Plateau to British Columbia, Alaska, Montana, and Wyoming. Up to two feet tall, it has broad, dark green, deeply veined leaves and fragrant flowers with brownish purple sepals and petals and a contrasting white lip with purple veining. *C. montanum* is on the CNPS "watch" list (List 4). It is occasionally available from specialty suppliers.

Epipactis gigantea, stream orchid
Prefers: low to mid-elevation, light shade, wet places, good drainage
Accepts: sun with moisture
Blooms: late spring, summer
Stream orchid is found in seeps and wet meadows and along streambanks throughout much of California north to British Columbia and south and east to

Epipactis gigantea

Texas and Mexico. Our most robust native orchid, it often forms extensive colonies, multiplying freely by seed and forming dense clumps or thickets from closely branched rhizomes.

Epipactis gigantea. (DOREEN SMITH)

The erect shoots grow from one foot to over three feet tall and are neatly lined with lanceolate to widely elliptic, bright green, deeply veined leaves. Plants bloom from late spring to late summer. Showy, greenish yellow flowers with red-purple veins are carried above the leaves. The flowers are intricate in structure, with cupped sepals and petals and a hinged, boat-shaped lip. The plant dies back to the ground in late fall and is dormant in winter.

Stream orchid may be planted streamside or in other moist spots, in full sun or light shade. In shaded sites ordinary garden watering will suffice, while sunnier exposures necessitate more frequent watering to keep the roots constantly moist. Plants thrive in containers (ideally, large tubs) as long as they are kept moist and the clumps are divided when they become crowded.

Stream orchid tolerates a wide variety of soils, although drainage should be reasonably good. Unlike most other California orchids, it is grown and offered commercially by several specialty nurseries. The most commonly available stream orchid is *Epipactis* **'Serpentine Night'**, a plant with dark red or purplish leaves that is grown mostly for its attractive foliage. The flowers are light green with purplish veins on two-foot stems. This form was selected from the wild by Roger Raiche in The Cedars in Sonoma County. It makes a good container plant, and appreciates sun, reasonably well-drained soil, and regular watering in spring and summer.

Goodyera oblongifolia, rattlesnake plantain

Prefers: mid-elevation, shade, moisture, good drainage, acid soil
Blooms: summer

Rattlesnake plantain is an evergreen orchid, native to dry coniferous forests in decomposing leaf litter in the Cascade Ranges, north and central Sierra Nevada, southwestern San Francisco Bay Area, and Modoc Plateau to Alaska and Mexico. It often shares mossy hummocks with such gems of the forest as pipsissewas (*Chimaphila* spp.) and pyrolas.

Lanceolate to widely elliptic leaves are arranged in basal rosettes that nestle on the ground, often in tight clusters issuing from a branched rootstock. The leaves may be uniformly deep green, but more often have a white-striped midrib or white reticulations throughout. All forms have a beautiful, velvety leaf surface. From early to late summer a mature rosette produces a spike of odd, tiny, greenish white blossoms, complex and interesting when viewed closely.

Rattlesnake plantain is a good, though not easy, candidate for a shaded nook in the open garden if provided with loose, humus-rich soil and placed high enough

Goodyera oblongifolia

that water will not collect around the crown. It also can be grown in containers: if planted in a light mix, shaded, and allowed to dry out slightly between waterings, it will soon fill the pot with attractive rosettes that can be divided and given to friends. This is among the hardiest of California orchids, although coastal plants probably are less so than those from higher elevations. Unfortunately, only a few specialty nurseries offer it, and much of the material available commercially is still collected from the wild.

ENDNOTES

[1] James C. Hickman (ed.), *The Jepson Manual: Higher Plants of California* (Berkeley: University of California Press, 1993).

[2] Peggy Fiedler, *Rare Lilies of California* (Sacramento: California Native Plant Society, 1996).

[3] L. Watson and M. J. Dallwitz, "The Families of Flowering Plants: Descriptions, Illustrations, Identification, and Information Retrieval," December 2000 (http://biodiversity.uno.edu/delta/); R. F. Thorne, "Classification and Geography of the Flowering Plants," *Botanical Review* 58 (1992): 225–348; R. M. T. Dahlgren et al., *The Families of the Monocotyledons: Structure, Evolution, and Taxonomy* (Berlin: Springer-Verlag, 1985).

[4] For example, 67 percent of the state's native lilies (*Lilium* spp.), 65 percent of fritillaries, and 62 percent of erythroniums are rare or uncommon. Fiedler, *Rare Lilies,* p. 4.

[5] Watson and Dallwitz, "Families of Flowering Plants."

[6] Hickman (ed.), *Jepson Manual; Sunset Western Garden Book* (Menlo Park, Calif.: Sunset Publishing Corp., 2001).

[7] Watson and Dallwitz, "Families of Flowering Plants."

[8] P. J. Rudall et al., *Monocotyledons: Systematics and Evolution* (Kew, Eng.: Royal Botanic Garden, 1995); M. W. Chase et al., "Monocot Systematics: A Combined Analysis," ibid., pp. 685–730.

[9] Watson and Dallwitz, "Families of Flowering Plants."

[10] Rudall, *Monocotyledons.*

[11] Watson and Dallwitz, "Families of Flowering Plants." Dahlgren, *Families of the Monocotyledons,* places *Scoliopus* in the Trilliaceae.

[12] Watson and Dallwitz, "Families of Flowering Plants."

[13] Hickman (ed.), *Jepson Manual.*

[14] Watson and Dallwitz, "Families of Flowering Plants."

[15] Ibid.

[16] Dahlgren, *Families of the Monocotyledons.*

[17] USDA, Forest Service, and USDI, Bureau of Land Management, "Final supplemental EIS on managing of habitat for late successional and old-growth species within the range of the northern spotted owl," 1994.

[18] Watson and Dallwitz, "Families of Flowering Plants."

[19] Dahlgren, *Families of the Monocotyledons.*

[20] Watson and Dallwitz, "Families of Flowering Plants."

[21] R. F. Thorne, "Proposed New Realignments in the Angiosperms," *Nordic Journal of Botany* 3 (1983): 85–117.

[22] Thorne, "Classification and Geography of the Flowering Plants."

[23] Watson and Dallwitz, "Families of Flowering Plants," place *Streptopus* in the Uvulariaceae, along with *Clintonia*, *Disporum*, and *Scoliopus*.

[24] Ibid.

[25] Ibid.

[26] Ibid.

[27] Dahlgren, *Families of the Monocotyledons;* Chase, "Monocot Systematics"; Watson and Dallwitz, "Families of Flowering Plants."

[28] Fiedler, *Rare Lilies.*

[29] Ibid.

[30] Hickman (ed.), *Jepson Manual;* A. Cronquist, *The Evolution and Classification of Flowering Plants* (Bronx: New York Botanic Garden, 1988); Thorne, "Proposed New Realignments in the Angiosperms."

[31] Thorne, "Classification and Geography of the Flowering Plants"; Dahlgren, *Families of the Monocotyledons;* Watson and Dallwitz, "Families of Flowering Plants"; A. Takhtajan, *Diversity and Classification of Flowering Plants* (New York: Columbia University Press, 1997).

[32] Dahlgren, *Families of the Monocotyledons;* R. M. T. Dahlgren and H. T. Clifford, *The Monocotyledons: A Comparative Study* (London: Academic Press, 1982).

[33] The three genera are placed by some authors in their own family, the Themidaceae. Chase, "Monocot Systematics"; M. F. Fay and M. W. Chase, "Resurrection of Themidaceae for the Brodiaea Alliance and Circumscription of Alliaceae, Amaryllidaceae, and Agapanthoideae," *Taxon* 45 (1996): 441–51; Rudall, *Monocotyledons.*

[34] Hickman (ed.), *Jepson Manual.*

Muhlenbergia rigens (SAXON HOLT)

III. GRASSES AND GRASSLIKE PLANTS

Native grasses and grasslike plants add to the splendor of almost every area of the California landscape—the solemn grandeur of coastal redwood forests and giant sequoia groves, high Sierra peaks and meadows, the precipitous Big Sur coastline, chaparral-clothed mountainsides, deep, shady canyons, rushing river courses and creeks, humid delta sloughs, flowery desert valleys, parklike oak savannahs, and open wildflower fields and grasslands. A minor component in some natural areas, dominant in others, grasses and grasslike plants are inextricably intertwined with popular images of California.

Grasses and grasslike plants grow all over the world in many different climates, on varied topography, and in an astounding variety of forms, from tiny-leafed tussocks to towering forests of timber bamboo. Since prehistoric times, they have been used by humans in many ways—for food, shelter, clothing, and medicine, as well as in art and religion. Domestication of the grasses that became our grains, coupled with livestock that foraged on these plants, profoundly affected the course of human population growth and settlement. Agriculture, largely grass-based, paralleled—even created—civilization as we know it today.

Grasses and grasslike plants are integral members of nearly every native plant community in California. Although most of the grasses we see in grasslands today are introduced nonnative species, these panoramic sweeps are visually reminiscent of former native stands. Both in the landscape and in the intimate detail of form, movement, sound, and smell, native grasses add richness and poetry to our lives.

When used creatively, native grasses and grasslike plants evoke the beauty of California even in the small scale of the residential garden. Nothing creates a naturalistic effect as convincingly and as easily. Because local native species are adapted to the region's climate, they are often easy in gardens. Native grasses and grass-like plants attract native animals, and they encourage native soil microorganisms to flourish. They also improve soil tilth as their voluminous root systems infiltrate, die, and regenerate. The diversity of habitats these plants enjoy in the wild provides ample suggestions for their use in nearly any garden setting, whether pond margin, shrub or perennial border, grassy meadow, woodland, or rock garden.

Grasses and grasslike plants might be used in drifts of a single species or several similar species to create visual continuity or as a solitary accent to contrast with and soften the harder forms of shrubs and perennials or architectural features. Some can be blended together for elegant and subtle contrasts of form or color. While the dominant form of grasses and grasslike plants is linear, often arching and fountainlike in habit, there are many variations, from bold, stiff, and erect to delicate, lax, and sensuous. The color palette offers lime green, emerald, bronze, rust, salmon, golden, tan, beige, dusty white, gray, or blue, and most grasses change color with the seasons.

Indoors too, the foliage and inflorescences of grasses and grasslike plants are valued for arrangements, wreaths, and other crafts. Whether chosen for their architectural or abstract form, their suggestion of nature, or their symbolic representation of abundance and fecundity, native California grasses and grasslike plants offer a wide range of material.

These plants also are utilitarian workhorses of the landscape. Binding soil is one of the functions many of them do best. Some form spreading mats of underground roots or creeping surface stems that, along with their upright stems and leaves, resist wind and water. Some are colonizers, establishing themselves on anaerobic muds, sterile sands, gravels, and rocky soils of many types, including serpentine, where few other plants will grow.

Many native grasses and grasslike plants serve well as low-maintenance groundcovers and can substitute for an ornamental lawn. Grasses that have shown promise for this purpose include purple needlegrass (*Nassella pulchra*), California oatgrass (*Danthonia californica*), Diego bent grass (*Agrostis pallens*), saltgrass (*Distichlis spicata*), blue grama (*Bouteloua gracilis*), and red fescue (*Festuca rubra*). Sedges such as *Carex globosa*, *C. pansa*, and *C. praegracilis* also have been used in this way, as has spikerush (*Eleocharis parishii*). Many more need to be tried, and tried in diverse climates.

Native grasses and grasslike plants that occur naturally on public or private property can be the focus of a restoration or preservation effort. As the natural landscape is increasingly fragmented and degraded by human activity, it becomes more important for landowners to preserve and enhance local forms of plants that occur on their property or nearby. Since even widespread species may differ significantly from one locale to another, preservation of local gene pools is critical. Any landowner can have a positive impact on the long-term survival of these plants.

To the uninitiated, grasses and grasslike plants may look more or less alike, but they represent at least three distinct families: the grass family, Poaceae; the sedge family, Cyperaceae; and the rush family, Juncaceae. The cattail family, Typhaceae, is another group of grasslike plants that is closely related to the sedges and rushes.[1]

Grasses are perennial or annual tufted herbs, usually growing from rhizomes or stolons and found in a variety of dry to moist habitats. They have erect to creeping stems with hollow internodes and jointed nodes, linear leaves, and inflorescences with small, inconspicuous flowers. Sedges are perennial or sometimes annual grasslike herbs growing from creeping rhizomes, usually in damp or marshy habitats. They have solid, three-sided stems, linear grasslike leaves, and inflorescences with inconspicuous flowers. Rushes are annual or perennial grasslike herbs that grow from creeping rhizomes in damp habitats. They have flat, irislike or narrow, cylindrical leaves and inflorescences with inconspicuous flowers. Cattails are perennial aquatic or semiaquatic herbs that grow from rhizomes, often forming large colonies in shallow water. They have linear, thick, flattened leaves and inconspicuous sessile flowers on tall, striking inflorescences.

Individual flowers of grasses and grasslike plants typically are small and not conspicuous, but the inflorescences in which they are combined can be quite showy. Grass inflorescences are highly varied, from narrow, erect or nodding spikes to fluffy clusters or dense to finely branched, ethereal plumes. At flowering, fertile anthers and pistils projecting from the floral bracts are lovely when observed closely. However, it is usually the bracts (technically glumes, lemma, and palea in grasses), or the enclosed seeds, that are most noticeable and that carry the "floral" show. Bracts may be covered with hairs or projections that add to the radiant quality of the inflorescence, especially when backlit by the sun.

Sedge flowers are often arranged spirally on the inflorescence branches, forming conelike clusters or erect or drooping catkins that top the stems or project from the sides. Some are showy, especially when the anthers and pistils are fertile and even for many months afterward; in seed they may resemble finely braided beadwork. The most common color of sedge inflorescences is brown, often deep brown, with overtones of burgundy, mahogany, russet, or black. Other sedge inflorescences are inconspicuous, and the ornamental value of the plant is in its foliage.

Rush flowers usually are clustered in small clumps or airy little clusters, typically emerging from the upper part of wiry or spiky stems. A marvel of detail in flower and seed, their color is similar to that of the sedges—variations of brown, tan, or green. Some, such as *Juncus lesueurii*, have showy flowers, more colorful than those of the sedges.

Cattail flowers are borne in inflorescences that resemble sausages or cats' tails. In some cattails the male or pollen-bearing flower spike is joined to the lower, female spike. In others there is a gap on the stem between them. The bur-reeds, closely related to cattails, bear spherical male flowers above spiny, round female flowers on a zigzag stem.

Stems and leaves are outstanding features of most grasses and grasslike plants. The greatest range in size, shape, and color probably occurs in the grasses, even though all are variations of a long, thin blade. Grass stems are hollow, with joints or nodes along the length, topped with flowers in inflorescences that vary with the species. In a few species, the stems are tall and woody, although this is the exception in California grasses. In many sedges and rushes the stems are the conspicuous part of the plant, while the leaves are short and inconspicuous. But some sedges have conspicuous leaves, especially in the genus *Carex*, and there is considerable variation in length, width, color, and whether or not the leaf is ridged or pleated. Sedge stems are mostly solid or partitioned. Cattail stems are erect and hard. Only grass stems are round and hollow with joints or nodes.

The leaves of some grasses and grasslike plants can be quite similar—long, slender, and pointed, with sheaths clasping the stems. Some sedges and rushes have thick leaves, sometimes with inner partitions. Cattail leaves are thick and somewhat spongy. Most grasses and grasslike plants have fibrous roots or heavy rootstocks; some have rhizomes or stolons by which plants spread vegetatively. Knowing whether the plants spread underground and how quickly they spread is important when using them in the garden, as the more invasive ones may need to be restricted to containers or used only in large, naturalistic landscapes.

Grasses and grasslike plants, especially sedges and rushes, can be difficult to identify from cut specimens, although they preserve well and thus provide many opportunities to attempt identification. Simple features such as the presence of nodes in grasses or the distinctive inflorescences in cattails and bur-reeds are helpful in distinguishing one group of plants from another.

GRASSES

Grasses are members of the Poaceae or grass family, which has 650 to 900 genera and about 10,000 species worldwide.[2] California's native grasses have adapted to an enormous range of climates and microclimates, but most of these are characterized by cool, wet winters and hot, dry summers. Two different though sometimes overlapping patterns of growth have emerged, based primarily on when the grass is actively growing.

Cool-season grasses start fresh growth with fall rains, flower in early spring to early summer, then set seed and go partially or fully dormant. These grasses tend to be from moderate-climate coastal plant communities where both winters and summers are mild. Warm-season grasses are mostly dormant through the winter, develop most of their growth in summer, and flower in mid-summer to late summer. These tend to be from interior areas of California, the mountains, Great Basin, or deserts. Most warm-season grasses do best with heat during the growing season, and many tolerate considerable winter cold.

For both cool-season and warm-season grasses, the growing season is about six months long. In the garden occasional summer watering will extend or reactivate the growing season of some cool-season grasses, but warm-season grasses seldom grow actively in winter, regardless of climate or watering regime. Some people find grasses unattractive in their dormant phase, but when used creatively dormant or dried grasses do add color and texture to the garden, and most can be tidied up to a level acceptable even to fastidious gardeners.

Some grasses take several years to develop into good specimens, but many native grasses are rapid growers, and these may need dividing or renewal after a few years. Divisions are best done in late fall or winter for most species, making sure that the pieces are kept moist until new growth is established. Renewal may take the form of cutting to the ground, mowing, vigorous raking, combing (long- or short-handled clawlike cultivators or garden hooks are good for this), or other grooming technique as best suits the species, but most grasses benefit from occasional removal of dead thatch that builds up within the clump or colony. The gardener can serve as a substitute for nature's frequent fires or for grazing and trampling by herds of wild elk in times past, both of which removed excess growth and dead residue from the plants.

In an increasing number of communities throughout California, regulations concerned with fire hazard may require annual cutting back of grasses. This usually applies to large, meadow-type plantings rather than specimen plants. Gardeners in these areas will need to pay attention to species selection and flowering times, both for grasses and for any associate species, but attractive seasonal landscape effects are still achievable. Mowers set high, or line or blade trimmers, can be used for large areas, while hand cutting works best in more intimate spaces. Most grasses respond well to being cut back to a few inches above the crown, and if done after most growth and flowering are completed, this should have no negative effects on the plants. Some species may make a new showing of fresh growth soon after cutting back, especially if they are watered, although the second wave will not be as showy as in the primary growth period.

Propagating native grasses is not difficult. Most grasses can be propagated readily from seed, by division, or from cuttings of rhizomes or stolons. Seeds should be sown in late summer or fall for cool-season grasses and in spring for warm-season types, although fall sowing of warm-season grasses may work as well. Sow seeds thickly in pots or flats in potting mix. Firm the mix, then place seeds on top and cover them to twice their thickness by sieving soil over them and compressing it lightly. Soil should be kept moist until seeds have germinated and young plants are an inch or two inches tall. Individual plants then may be pricked out and placed into tubes, liners, or pots five to seven inches deep. Seeds also can be sown directly into tubes or one-gallon cans. The fewer times seedlings are transplanted, the better.

When plants have developed a good root system, and when rains have wet the ground where they will be planted out, they can be moved from tubes or pots to the open garden. Watering the first year will aid establishment, and fall plantings will need less supplemental irrigation than those placed out in spring. Plants grown outside their native habitats may need some irrigation even when mature.

With a good start, many grasses will bloom and set seed the first year. After seeds have ripened (about six weeks after flowering), they may be collected, dried, and stored in a cool, dry place for future planting.

Cuttings from sections of rhizomes or stolons can be started in a variety of mixes, such as sharp sand and peat, peat and vermiculite, or perlite and a light potting soil. Keep the mix moist until tops and roots have formed, then move them to pots for growing on to a size that can be planted out in the garden (when roots reach the container wall). Fall planting is best for most perennial native grasses, but with consistent irrigation they may be planted out at any time that plants are actively growing and summer heat is not too intense.

Direct seeding of native grasses onto prepared plots of open ground has had mixed success. For best results, seed should be pressed into the soil, covered slightly, and watered regularly until young plants are established. Direct sowing onto soil occupied by other plants generally is ineffective, as few seedling plants will compete with established vegetation. Controlling competition from weeds and other plants is the most challenging aspect of establishing native grasses in the landscape.

Growing Grasses

There are hundreds of creative and beautiful ways to use native grasses in the home landscape, but perhaps nothing is more suggestive of California's magnificent floral heritage than a flowery meadow. Experimentation with various species is encouraged, but it is important to combine plants of similar cultural needs, compatible scale, and sufficient variety to ensure a long season of interest.

A good way to lessen the impact of foot traffic through a meadow garden is to install an informal pathway through the meadow area using flat stones. The stones are softened and somewhat disguised by the billowing plants as they mature, yet they serve to guide footsteps through the area without compacting the soil or damaging the plants. Some possible plant combinations for grassy meadows are suggested below.

Dry meadow. A dry, sunny meadow could combine drier-growing grasses with native perennials, bulbs, a shrub or two, and annuals that thrive in sunny locations with little or no supplemental irrigation. A pleasing effect can be created by combining grasses such as nodding needlegrass (*Nassella cernua*), purple needlegrass (*N. pulchra*), foothill needlegrass (*N. lepida*), cane bluestem (*Bothriochloa barbinodis*), melic or oniongrass (*Melica californica* or *M. torreyana*), Junegrass (*Koeleria macrantha*), and muhly (*Muhlenbergia rigens*). Integrating mixes of cool-season and warm-season grasses, especially those that retain some color other than tan, such as *Leymus condensatus* 'Canyon Prince', or adding sedges that hold their green into summer months, such as *Carex globosa*, would extend visual interest into the dry season. Such combinations also accentuate the tapestry effect for which natural grasslands are admired.

These grassland plants could be combined attractively with sanicles (*Sanicula* spp.), various species of *Lomatium*, blue-eyed grass (*Sisyrinchium bellum*), *Brodiaea*, *Dichelostemma*, or *Triteleia* species, mariposa lilies (*Calochortus catalinae*, *C. luteus*, *C. venustus*), soap plant (*Chlorogalum pomeridianum*), buttercups (*Ranunculus* spp.), checkerblooms (*Sidalcea* spp.), wild buckwheats (*Eriogonum* spp.), coyote-mint (*Monardella villosa*), and California fuchsia (*Epilobium canum*). A few annuals to

Grasses in meadow garden. (CAROL BORNSTEIN)

add color and seasonal interest in early years of establishment might include some of the many species of *Clarkia*, annual lupine such as *Lupinus nanus*, California poppies (*Eschscholzia* spp.), *Phacelia tanacetifolia*, large-flower linanthus (*Linanthus grandiflorus*), and common madia (*Madia elegans*). However, as bunchgrasses increase and spread, many annuals will gradually decline.

Dry meadows can be accentuated by massed or specimen shrubs such as monkeyflower (*Mimulus aurantiacus*), toyon (*Heteromeles arbutifolia*), manzanita (*Arctostaphylos* spp.), mountain-mahogany (*Cercocarpus betuloides*), island buckwheat (*Eriogonum arborescens*), or red shank (*Adenostoma sparsifolium*). One might emphasize plants native to the local region, or mix together plants from all over the state or the world. There are many exciting combinations waiting to be tried.

Desert garden. Combining grasses with xeric chaparral or desert shrubs, cacti, and succulents could create a more arid effect. The most common aesthetic deficiency of cactus and succulent gardens is that they lack the shrubs and grasses that occur naturally in desert climates. The hard, architectural features of many of the larger cacti, succulents, and xerophytic monocots (such as *Agave*, *Yucca*, and *Nolina*) are perfectly complemented by the elegant fluidity of grasses that thrive in arid climates.

Grasses in desert garden. (ROGER RAICHE)

The addition of purple or Wright three-awn (*Aristida purpurea* var. *purpurea* or *A. purpurea* var. *wrightii*), desert needlegrass (*Achnatherum speciosum*), the fascinating blue grama (*Bouteloua gracilis*), galleta (*Pleuraphis jamesii*), big galleta (*P. rigida*), and many of the species mentioned above for dry meadows will transform a collection of succulents into a garden of note. Further enriching the setting may be shrubs such as Mormon tea (*Ephedra* spp.), Apache plume (*Fallugia paradoxa*), Great Basin sagebrush (*Artemisia tridentata*) or other *Artemisia* species, arid buckwheats (*Eriogonum* spp.), desert ferns (*Cheilanthes* spp., *Notholaena californica*, *Astrolepis cochisensis*), matchweed (*Gutierrezia sarothrae*), *Stephanomeria cichoriacea*, sages (*Salvia* spp.), and beardtongues (*Penstemon* spp.).

For larger gardens, a focal shrub or small tree such as Utah juniper (*Juniperus osteosperma*), singleleaf pinyon pine (*Pinus monophylla*), manzanita (*Arctostaphylos* spp.), or ceanothus (*Ceanothus greggii*, *C. megacarpus*, *C. cuneatus* var. *cuneatus*) will combine to good effect. Open areas can be accentuated with seasonal annuals such as the smaller poppies (*Eschscholzia* spp. or *Stylomecon heterophylla*), lupines (*Lupinus* spp.), baby blue-eyes or fivespot (*Nemophila menziesii* or *N. maculata*), *Linanthus* species, red ribbons (*Clarkia concinna*), or Chinese houses (*Collinsia* spp.).

Moist meadow. A moister meadow garden could simulate a moist mountain meadow, a coastal prairie, or a damp woodland clearing. Montane meadows may be best left to specialists, as many alpine plants require special care, but pleasing low-elevation grasslands are relatively easy to reproduce. The moister meadows do best in central to northern California, especially in coastal mountains and canyons. Hotter interior or southern locations will require some experimentation to find a plant palette that works, or cultural conditions will need to be modified to increase chances of success. For best results, depending on climate, soil, and other conditions, water should be applied every two to four weeks during the dry season, and more frequently would favor some plants.

Because of the extra moisture and the longer-lasting lushness and color in these settings, moist meadows are easy to integrate into traditional landscapes and are especially appropriate near residences. Many moisture-tolerant grasses also are excellent candidates for inclusion in traditional border schemes.

Moisture-tolerant grasses include fescues (*Festuca rubra*, *F. californica*), reedgrasses (*Calamagrostis foliosa*, *C. nutkaensis*, *C. ophitidis*), hairgrasses (*Deschampsia cespitosa* ssp. *holciformis*, *D. elongata*), satintail (*Imperata brevifolia*), and panicgrass (*Panicum acuminatum*). Many sedges and rushes combine well with these grasses, although some of them will need more irrigation.

A large number of native shrubs, perennials, bulbs, and annuals also combine to excellent effect in moist meadow gardens. Some of the possibilities are moisture-tolerant bulbs such as some wild onions (*Allium unifolium*, *A. hyalinum*), *Brodiaea terrestris*, *Triteleia ixioides*, *T. hyacinthina*, or *T. peduncularis*, large-flowered star tulip (*Calochortus uniflorus*), *Odontostomum hartwegii*, soap plant (*Chlorogalum pomeridianum*), camas (*Camassia quamash*), and Fremont's zigadene (*Zigadenus fremontii*). These bulbs, however, prefer a dry summer dormancy, and they may rot if watered in late summer months.

Herbaceous plants for the moister meadow might include buttercups (*Ranunculus* spp.), checkerbloom (*Sidalcea malviflora*), sun cup (*Camissonia ovata*), some Pacific Coast irises (*Iris douglasiana* and others), yellow-eyed grass (*Sisyrinchium californicum*), wild strawberry (*Fragaria* spp.), *Horkelia* spp., *Geranium californicum*, *Helenium bolanderi*, marsh lupine (*Lupinus polyphyllus*), yampah (*Perideridia* spp.), or sea pink (*Armeria maritima*). For excitement add some annuals such as meadowfoam (*Limnanthes douglasii*), farewell-to-spring (*Clarkia amoena*, *C. rubicunda*), *Sidalcea calycosa*, large-flower linanthus (*Linanthus grandiflorus*), lupine (*Lupinus* spp.), and centaury (*Centaurium muehlenbergii*).

Moisture-tolerant shrubs that provide contrast, scale, and background might include bearberry (*Arctostaphylos uva-ursi*), coffeeberry (*Rhamnus californica*), Carmel ceanothus (*Ceanothus griseus*), dwarf coyote brush (*Baccharis pilularis*), barberry (*Berberis* spp.), woolly sunflower (*Eriophyllum* spp.), or red-flowering currant (*Ribes sanguineum* var. *glutinosum*).

Grasses for the Garden

Following are descriptions of some California grasses that are most useful for gardens. Omission of a species does not imply lack of horticultural value. Many grasses have yet to be tried in gardens.

Achnatherum parishii, Parish's needlegrass
Prefers: mid-elevation, dry, sun, good drainage
Accepts: light shade, some irrigation
Blooms: spring, early summer
Parish's needlegrass is a stout, somewhat stiff bunchgrass that forms loose clumps of flat, narrow leaves and bold, bristly spikes on stems two or more feet tall. Native to dry, rocky slopes, shrublands, and pinyon-juniper woodlands from the southern Sierra Nevada into the southern California desert mountains and Baja California, this cool-season grass is valuable as an accent in arid landscapes. It does best in well-drained soil in full sun, but tolerates light shade and supplemental water. Easy from seed, it self-sows readily, and it might serve as a rough bunchgrass turf in arid areas.

Achnatherum speciosum, desert needlegrass
Prefers: mid-elevation, dry, full sun, good drainage
Accepts: some irrigation
Blooms: spring, early summer
Desert needlegrass is one of California's most exquisite tightly clumping bunchgrasses. It forms an elegant, fountainlike display of fine, thin blades tightly packed into a one- to two-foot-tall clump, the blades emerging lime green when fresh, drying to golden tan. Rising above the foliage clump are stems to two feet tall terminating in a narrow, four- to six-inch-long brushlike spike covered with fine hairs.

   This lovely cool-season grass is found on rocky slopes and in arid canyons and washes of the southern California deserts to Colorado, Arizona, New Mexico, Mexico, and South America. It makes a fine addition to a rock garden or desert garden planting. Best grown in the open with excellent drainage and full sun, it tolerates some summer irrigation. Clumps do not renew well, so it is best to divide them in late winter or allow plants to self-sow.

Achnatherum thurberianum, Thurber's needlegrass
Prefers: mid- to high elevation, dry, sun, good drainage
Blooms: spring, early summer
Thurber's needlegrass is a seldom cultivated but attractive clumping grass from sagebrush shrublands of the Great Basin and arid mountains of southern California and much of the West. It forms fine-textured clumps of foliage to about one foot across, from which two- to three-foot spikes emerge, giving a strong vertical look that is softened by the branched inflorescence. In summer the plant develops a tan coloration, providing a nice accent for sunny, gravelly sites. Little is known about its culture, but it should be tried more often, especially in arid gardens.

Agrostis hallii, Hall's bent grass
Prefers: mid-elevation, sun or light shade, some moisture
Blooms: mid-summer
Hall's bent grass is a delicate grass, creating a soft groundcover effect. Its stems are slender, from one foot to two feet tall, with soft, arching, flat leaves one-eighth inch wide by six inches long spaced out along the stalks. Plants spread underground by rhizomes and make fine-textured colonies in light shade under trees or shrubs or in full sun in open grassland. It is especially useful as a filler among larger plants, as a groundcover, or as a soil binder on slopes. Sharing growth habits of both cool- and warm-season grasses, the foliage stays green into early summer when other native grasses have started to turn brown. Continued summer growth is enhanced by supplemental watering. The airy, green inflorescence is attractive but not showy.

Native to open oak woodland, coniferous forest, and open grassland in central and northwestern coastal California and Oregon, this grass makes a fine addition to the naturalistic landscape. Propagated from rooted sections or pieces of rhizome, it could be tried as a lawn substitute.

Agrostis pallens, Diego bent grass
Prefers: mid-elevation, sun or part shade, dry, good drainage
Accepts: some drought, some irrigation
Blooms: spring, early summer

Diego bent grass is similar to *Agrostis hallii*, but with even more finely textured foliage, tighter growth habit, and greater tolerance of dryness. It is native to open meadows, woodlands, and forests at mid-elevations in western California and the Great Basin to British Columbia. An attractive grass, it is especially useful as a groundcover under shrubs or scattered trees. It is excellent for meadow effects, mowed or unmowed.

Considered one of the best native grasses for a relatively drought-tolerant lawn substitute, Diego bent grass can be directly seeded on open, weed-free ground, watered until established, and mowed several times in summer. Without supplemental water it goes partially dormant in summer, but with a thorough watering every three or four weeks, coupled with mowing, it will produce a deep green turf of excellent quality. The attractive foliage makes this grass useful in container plantings. It is easy from seed or by division.

Agrostis scabra, ticklegrass
Prefers: mid- to high elevation, sun, regular water
Blooms: mid- to late summer
Ticklegrass is fine-textured, with stems one foot to two feet tall, topped by a remarkably delicate and airy spray of pinkish flowers that shimmer in sunlight. It grows at mid- to high elevations on open roadsides and in meadows and coniferous forests in the Klamath and North Coast Ranges and the Sierra Nevada east to Mono and Owens Valleys.

This warm-season grass appreciates regular summer irrigation. It forms dense tufts but also reseeds readily, making it useful as a groundcover or meadow plant. The delicacy of the inflorescence suggests backlighting or close display, perhaps in containers. Ticklegrass is easy from seed.

Aristida purpurea var. *purpurea*, purple three-awn
Prefers: low to mid-elevation, sun, dry, good drainage
Accepts: drought, poor soils
Blooms: mid- to late summer
Purple three-awn is an attractive grass native to dry slopes and shrublands at low to mid-elevations in the South Coast Ranges and desert mountains to northern Mexico. It has showy flower spikes on stems one foot to two feet tall and narrow foliage that forms fine, dense clumps from three to eight inches across. This warm-season grass produces erect and arching flower spikes in summer that have a purple cast when fresh and turn tan when dry. The long triple awns (stiff projections)

Aristida purpurea. (STEPHEN INGRAM)

produced by each flower in the spike combine to create an inflorescence that is dramatic in sunlight.

The small scale of this grass and its bristly look, great drought tolerance, and ability to grow in poor, stony soils suggests its use in a dry meadow or as a complement to shrubs, perennials, or cacti and succulents in the arid garden. It is easy from seed.

Aristida purpurea var. *wrightii*, Wright three-awn
Prefers: mid-elevation, dry, sun, heat, excellent drainage
Blooms: mid- to late summer
Wright three-awn is similar to *Aristida purpurea* var. *purpurea*, but shorter and without the purplish cast to the fresh flower spikes. The spikes are tan-colored, usually rising about a foot above the low clump of fine foliage. Even in late summer the clumps maintain an attractive golden green in the center, fading to beige-tan toward the leaf tips.

This grass is excellent in combination with smaller succulents and desert plants, where it will self-sow, forming a lovely haze of fine foliage and decorative bristly awns that last into winter. It is especially striking against the hard forms of plants such as prickly-pear cactus (*Opuntia* spp.) or *Agave* species or scattered among boulders in a hot, dry landscape. It is native to sandy or rocky slopes, plains, and shrublands of the Mojave Desert and the Peninsular Ranges to northern Mexico. It requires excellent drainage.

Bothriochloa barbinodis, cane bluestem
Prefers: mid-elevation, full sun, dry, good drainage
Accepts: some irrigation, heavy clay soil
Blooms: summer

Cane bluestem is a dramatic grass, two to four feet tall, with prominent, silvery white flowerheads of great beauty. Native to dry slopes in southern California, Texas, and Mexico, this plant shares characteristics of both cool- and warm-season grasses, provided water is available. The clump is leafy, rather stiff, and slightly coarse up close, and provides a strong accent when grown among finer or smaller grasses. The bold, cottony flowerheads, conspicuous at a considerable distance and produced throughout the year in mild climates if watered, are the main reason to grow this grass.

One of the most charming and adaptable of ornamental grasses, cane bluestem, also known as mosquito grass or eyebrow grass, can tolerate heavy clay soil, and plants persist for many years with minimal care. It requires full sun and may self-sow in moist soils, but is seldom weedy. Wonderful as an accent or in mass, it adds contrasting form and a long season of interest to meadows or shrub margins. When backlit by the sun, the stems become sparkling frosted wands; against a dark background they appear as fiery torches. It is grown from seed, which must be cleaned of its hairy plumes, or by division.

Bouteloua curtipendula, side-oats grama
Prefers: mid-elevation, sun, heat, dry, good drainage
Accepts: moderate water
Blooms: late spring, summer

Side-oats grama is an attractive mid-sized grass with flowering spikes one foot to three feet tall and spreading basal clumps of fine foliage only a few inches tall. The tall, wiry spikes of this warm-season grass are produced in late spring and summer. Each spike is ornamented with up to sixty little flower clusters that resemble grains of oats that dangle or jut out to one side of the spike, creating an elegant effect that is accentuated by fading light or a breeze. When in true flower, each little side cluster produces a delightful display of protruding orange anthers, a subtle detail that is long remembered.

This grass is native to dry, rocky slopes, sandy to rocky drainages, scrub, and woodland throughout much of California to southern Canada, the eastern United States, and South America. With the strong vertical quality of its spikes, it is particularly useful as an accent, and it is attractive in flower arrangements. It does well in heat and with infrequent water and needs good drainage. It is grown from seed or by division.

Bouteloua gracilis, blue grama
Prefers: mid-elevation, sun, heat, dry, good drainage
Accepts: drought, light shade
Blooms: late spring, summer
Blue grama is a versatile and fascinating small grass native to rocky slopes, flats, drainages, scrub woodland, and pine forests at mid-elevations in arid mountains of southern California and beyond. It forms low, dense, slowly spreading clumps of narrow, grayish green leaves about one inch to three inches tall. In late spring and summer it produces graceful spikes, usually under a foot tall, though occasionally to two feet. Jutting from the side of each spike are jaunty brush- or comblike, purplish structures about two inches long, the entire inflorescence giving the effect of a pole with flags or banners snapping in the breeze. When in true flower, yellow anthers dangle from these purplish banners, adding yet more intricate beauty.

Bouteloua gracilis. (CAROL BORNSTEIN)

This warm-season, sod-forming grass makes a good groundcover needing little water and minimum care, doing fine in full sun or light shade but requiring good drainage. It is a fine lawn substitute when mowed about once a month during the growing season; left unmowed, it makes an informal soft meadow, where its flaglike flowers seem to float above the turf. Used as the dominant groundcover in a meadow, it forms an excellent carpet in which to feature bolder grasses, perennials, or shrubs. Even in winter, when the foliage is predominantly tan, some green shows through, and the plant maintains its tidy appearance. It is grown easily from seed or by division.

Calamagrostis foliosa, leafy reedgrass

Prefers: mid-elevation, cool, coastal habitats, full sun to part sun, regular water, good drainage
Accepts: some dryness
Blooms: spring, early summer

Arguably California's most beautiful native grass, this cool-season grower is native to wind-whipped coastal bluffs, scrub, and forests of northern California. Its distribution is limited, however, and it is on the CNPS "watch" list (List 4). Nursery-grown plants and seeds are readily available.

Leafy reedgrass makes the transition to gardens with great style and beauty. The twelve- to eighteen-inch-wide foliage clumps consist of flat, greenish gray leaves, about one-quarter inch wide in a flattened hemisphere. In spring, six-inch-long, narrow, grayish tan plumes tinged with purple emerge abundantly on one-foot-tall stems from the foliage clump, like a dazzling fireworks display. These exuberant plumes are produced right to the ground if not crowded, even downhill if on a slope.

This is an excellent plant as a specimen or in mass, in the rock garden or perennial border, above a wall or on a bank, or in a meadow or woodland margin. It does best in cool, central to northern California coastal climates, and is most gloriously grown in good soil with good drainage, routine water, and full sun.

Plants usually need to be renewed every two or three years by cutting back and removing any build-up of residue, or old plants

Calamagrostis foliosa.
(SAXON HOLT)

can be removed and nearby self-sown seedlings allowed to take over. This showy grass could even be used as an annual bedding plant, with new plants placed out in late summer for the next season's spring and summer display. Plants grown drier tend to last longer, but are not as effusive in foliage and flower. Leafy reedgrass grows easily from seed, less easily by division.

Calamagrostis koelerioides, tufted pine grass

Prefers: mid- to high elevation, sun, some water, excellent drainage
Blooms: summer

This is another attractive reedgrass, forming casual clumps or slowly spreading colonies of quarter-inch-wide foliage about twelve to eighteen inches long and vertical spikes two to three feet tall topped with soft, narrow plumes. The leaves are soft green, aging to a mix of gold, olive, and bronze. This grass is lovely in the wild, where it occurs in rugged terrain in chaparral and upper ridgelines of mountains throughout much of California. It is also found in Idaho, Wyoming, and Baja California.

Tufted pine grass does not thrive in low-elevation gardens, but is a choice plant for the grass enthusiast in mountain areas. It requires excellent drainage. It is propagated from seed or by division.

Calamagrostis nutkaensis, Pacific reedgrass

Prefers: low elevation, sun or light shade, cool temperatures, coastal habitats, regular water, good drainage
Accepts: heavy wet soil, some shade
Blooms: spring, early summer

Pacific reedgrass is a tough, robust bunchgrass two to four feet tall. Its bold, visually commanding structure is created by the substantial clump of erect, broad, flat, half-inch-wide leaves, from which strong, attractive flower stalks emerge in long, loose, nar-

Calamagrostis nutkaensis. (CAROL BORNSTEIN)

row plumes. Native to moist, foggy beaches, dunes, and coastal conifer forests of the North and Central Coast Ranges and the San Francisco Bay Area, it is a cool-season grower, but can be kept green year round with occasional watering.

This grass is effective as a specimen, in mass, or even as a low hedge. It combines nicely with medium-sized coastal shrubs such as *Rhamnus*, *Ceanothus*, *Arctostaphylos*, *Vaccinium*, *Rhododendron*, or *Myrica* species. In colder climates it provides attractive winter color, including yellows and oranges. It needs good drainage.

Pacific reedgrass displays two different growth habits, depending on the conditions under which it grows. Plants from open coastal meadows or bluffs tend to be clump-forming, and in shade the lower stems may root and form new plants. Plants from coastal conifer woodlands sometimes have the genetic capacity to spread underground, as well as to root from surface stems. These are naturally better suited for groundcover use, but may be difficult to control in smaller gardens. Both types tolerate heavy, wet soils and some shade. Plants grown in open shade tend to become colonial and hummocky, but this can be controlled, if desired, by annual cutting to the ground.

Some plants may develop purple coloration in the leaves. This is thought to be caused by environmental stresses (water, nutrients, or temperature), and in most cases is not unattractive. Plants are easily grown from seed or by division of layered stems.

A selection by Roger Raiche from the King Range in Mendocino County and called "The King" is noted for its exceptionally broad, glossy, bright green leaves. This big, husky, almost tropical-looking grass merits more attention. It is truly handsome and garden-worthy.

Calamagrostis ophitidis, serpentine reedgrass
Prefers: mid- to high elevation, serpentine soil, shade, some moisture, excellent
   drainage
Blooms: spring, early summer
Serpentine reedgrass is a handsome bunchgrass two to three feet tall and twelve to eighteen inches wide with quarter-inch-wide, erect blades and straight, narrow plume inflorescences that look like sparklers. A cool-season grower that can be prolonged into summer with extra water, it is notable not only for the fine, tidy form of the clump and flowering spikes, but for the sublime range of colors it displays as the seasons progress. In late winter and spring the foliage is pale green, but as the flowering spikes emerge the entire plant picks up shades of gold, bronze, olive, and salmon, glowing in the low light of morning or evening.

Serpentine reedgrass thrives in poor, gravelly or rocky, well-drained sites, where it is long-lived; it is shorter-lived on clay soils. Seldom grown, its success in cultivation outside central California is uncertain, but its beauty recommends far greater experimentation. It is excellent as an accent or in drifts in meadows, rock gardens, or shrub borders. It is propagated from seed or by division and performs best in part shade inland or full sun along the coast.

An uncommon native of serpentine grasslands and chaparral of the North Coast Ranges and the San Francisco Bay Area, it is on the CNPS "watch" list (List 4). Seeds and plants are available from specialty nurseries and seed suppliers.

Calamagrostis purpurascens, purple reedgrass
Prefers: high elevation, sandy or rocky soils, full sun, routine water, excellent
   drainage
Accepts: low elevation if grown on rocky mounds, some dryness
Blooms: summer

Purple reedgrass forms tight clumps of narrow, grayish green foliage about one foot tall with many tidy spikes that appear to be shooting up out of the clump. The narrow plumes initially have a purplish cast but quickly fade to beige, and they persist for most of the season. Native to alpine and subalpine dry, rocky slopes and moist meadows in the Cascade, White, and Inyo Mountains of California to Colorado, Montana, Wyoming, Alaska, and Siberia, it has proven a challenge to grow at low elevations.

Best for rock gardens with well-drained, gravelly soils and regular irrigation, it requires extra attention to remain vigorous. Probably of great utility in high-elevation gardens, its beauty makes it worth trying elsewhere. It is grown from seed or by division.

Calamagrostis rubescens, pine grass
Prefers: mid-elevation, sun, sterile soils, some summer water, good drainage
Blooms: spring, early summer

Pine grass is an attractive, spreading grass that forms a loose, hummocky groundcover in light shade. The leaves are about one-quarter inch wide and twelve inches long and arch loosely downward, forming a finely textured mix of pale green, tan, beige, and russet tones. While pine grass seldom flowers in shade, in open sun it produces occasional spikes of loose plumes to about two feet tall. However, it is the fine, tussocky quality of the foliage that is of most interest in the garden.

Native to dry woods and chaparral in the North Coast Ranges to southwestern Canada and the Rocky Mountains, especially on north-facing slopes in sterile soils, pine grass can be used effectively to unify mixed plantings, cover bare soil under shrubs or trees, form a transition between open meadow and forest, or provide cover for native bulbs or ephemeral perennials.

A cool-season grass, pine grass can be kept actively growing well into summer with extra water, though some may prefer the beige to golden-bronze tones accentuated by growing it dry. It does best with some summer water and needs good drainage. It is grown by division and, less commonly, from seed.

Danthonia californica, California oatgrass
Prefers: low to mid-elevation, sun, cool temperatures, coastal habitats, some summer water
Accepts: any soil, moderate foot traffic, some drought along coast
Blooms: spring, early summer
California oatgrass makes soft cushions two to six inches high and up to a foot

Danthonia californica

wide. The leaves are flat, bright deep green on the surface and grayish green below, frequently with hairs, and remain green into early summer. The flowering spikes extend about a foot above the foliage and consist of three to five open, spreading branches tipped by a cluster of bractlike flowers. The spike initially sprays upward, then bends down to nearly horizontal when the seeds are ripe. After ripening, it conveniently detaches from the clump, making cleanup easy with a good raking.

Native to open meadows and forests in many parts of California and the West to South America, plants in cultivation seem to do best in coastal regions of central and northern parts of the state. California oatgrass prefers occasional summer water, but can be grown dry in cooler coastal regions. Even when dry, plants retain some green in the center of the clump, although the outer length of the leaf will be tan with bronze flecks. This is an excellent colonizing grass for harsh, sterile sites, including disturbed soils, gravels, clays, sandstone, volcanics, and serpentine.

California oatgrass takes trampling well, which only makes the plants lower and flatter, so it has potential for areas of moderate foot traffic. It is useful for mowed or unmowed turf, as plantings fill in with age. It is good in meadows, especially as a filler among larger grasses or in openings in shrub plantings, and for erosion control on difficult sites. It tolerates part shade. It self-sows and is easy from seed, but seeds normally do not germinate for two years. It can be grown by division.

Danthonia californica* var. *americana has foliage that is more hairy, and it is often found at higher elevations than are plants of the species, although the two are sometimes found together. It is considered less vigorous in cultivation, but should be useful in cooler regions. It needs a sunny spot and good drainage.

Danthonia intermedia, timber oatgrass

Prefers: mid- to high elevation, sun, dry to moist, good drainage
Accepts: some dryness, some shade
Blooms: spring, summer

This is an attractive, compact clumping bunchgrass that produces dense, three- to six-inch-tall foliage, gray-green with tan-gold tips. The flower spikes, airy and under a foot tall, appear in spring at lower elevations and in summer in the mountains. Timber oatgrass is native to a wide range of habitats, including rocky outcrops, dry to moist meadows, and forest openings from mid- to high elevations in the Sierra Nevada and the Klamath and North Coast Ranges to Colorado, Utah, Montana, and South Dakota. This grass makes a fine addition to a rock garden with gravelly soil. Plants self-sow slowly and can be grown from seed or by division.

Deschampsia cespitosa ssp.
cespitosa, tufted hairgrass

Prefers: low to high elevation, cool, coastal habitats, year-round moisture, sun or part shade

Blooms: early spring, summer

Tufted hairgrass is a beautiful dark green bunchgrass that is highly variable within its enormous natural range in high-elevation mountains of both the northern and southern hemispheres. It is found at lower elevations on serpentine in the Coast Ranges. Native to moist meadows, streambanks, and bogs, it requires regular water year round to perform well and prefers a cool climate, such as that of coastal northern California.

Deschampsia cespitosa ssp. *cespitosa*. (ROBERT C. WEST)

Tufted hairgrass typically forms clumps about a foot tall, but can be twice that in rich soils. Each clump is composed of densely packed, narrow leaves about one-eighth inch wide, frequently inrolled and appearing threadlike. The exuberant, airy panicles extend two to three feet beyond the foliage on golden stems and are composed of hundreds of tiny flowers on fine, repeatedly branched branchlets, creating a shining, nearly transparent plume, especially when illuminated by the sun.

A cool-season grower, tufted hairgrass will remain green year round with adequate water, but develops tan leaf tips if allowed to dry out. Experience has shown that high-elevation plants do not thrive, or even survive, in low-elevation gardens. Plants found at low elevations in seepages in interior regions of the state are thought to be the best choices for cultivation in hotter areas. In coastal regions, this grass should be given a prominent place in the garden in sun or light shade. It is especially effective in mass plantings, as a soil binder on gentle slopes, or in irrigated meadow landscapes.

Tufted hairgrass self-sows prolifically in open, irrigated soils, so it should be mulched to control seedlings if spread is not desired. Even in the best sites, clumps are not long-lived, and after a few years it is best to remove them and allow new seedlings to take over. Plants are grown from seed or by division.

Deschampsia cespitosa
ssp. *holciformis*

Deschampsia cespitosa **ssp.** *holciformis*

Prefers: mid-elevation, coastal habitats, sun, regular water

Accepts: some salinity, some dryness along coast or in shade

Blooms: early spring, summer

This is a lower-growing subspecies of *Deschampsia cespitosa* that is native to coastal meadows and marshy areas in the North Coast Ranges to British Columbia. Not only is the foliage lower and stiffer than in the species, but plants flower later and the inflorescence is denser and frequently displayed at an angle to the clump. Reportedly tolerant of drier conditions, this probably is true only in coastal regions, except in shade. It tolerates some salinity.

The cultivar **'Jughandle'**, collected by David Amme, is notable for its stiff, sharp, almost yuccalike blades. It forms low, dense foliage clumps and showy inflorescences that preserve the low profile of the foliage. The clumps look like green stepping stones and will tolerate light foot traffic, but trampling will spoil the floral spikes.

Deschampsia danthonioides, annual hairgrass

Prefers: mid-elevation, sun, vernally moist conditions

Blooms: late spring, early summer

Annual hairgrass is an extremely fine-textured annual clumping grass with considerable horticultural potential. Forming clumps of foliage from four to eighteen inches tall, in late spring and early summer it is topped by numerous airy panicles of great delicacy that sway and shiver in the slightest breeze. Native to a wide range of western North America into South America, typically in open, vernally wet sites that dry out in summer, it is easy in cultivation.

Annual hairgrass self-sows prolifically but is probably most easily propagated by starting seeds in flats or pots in winter and setting out plugs of seedlings in early spring. In pots or beds of spring annuals, the hazy, diffuse plumes provide delightful counterpoint in texture, color, and motion and add a note of naturalism to a bedding display. As plants dry in summer, some seed can be saved for the next season and the rest removed to control self-sowing.

Deschampsia elongata, slender hairgrass

Prefers: mid- to high elevation, sun to part shade, moisture, good drainage

Blooms: late spring, summer

Slender hairgrass forms refined, tiny tufts of slender, pale yellow-green leaves, usually under four inches long, and elegant flower stalks one foot to four feet tall.

A delicate perennial grass, it tends to be short-lived. It seeds abundantly, but is easily pulled if it spreads into areas where it is not wanted. Native to a wide area of western North America into South America, it is found in damp meadows, along lakeshores, and in areas with moisture at least into late spring.

The vertical lines, airy plumes, and glowing yellow-green foliage suggest many potential uses in the garden, whether in containers of annuals or perennials, for tapestry meadow effects, or as subtle rock garden accents. Plugs set into a turf of the frosty-blue *Festuca rubra* 'Patrick's Point' or contrasting with the bronze stream orchid, *Epipactis* 'Serpentine Night', would create memorable color schemes. Seldom grown, it deserves much wider use. It is easy from seed or by division.

Distichlis spicata, saltgrass

Prefers: low to mid-elevation, sun, moisture
Accepts: salinity and alkalinity, some drought
Blooms: late summer

Saltgrass, native to salt marshes and alkaline or saline seepages and flats throughout much of North and South America, spreads rapidly from rhizomes and has potential as coarse turf in difficult sites. Its tolerance of heat, drought, and salty or alkaline water and soils and its tough constitution and resilience to foot traffic and grazing make this a useful plant for large-scale coverage, erosion control, and areas of hard use. A warm-season grower, it forms spreading mats of sprawling, gray-green leaves and small clusters of purplish flowers, although these are not showy.

This grass can be mowed, but it is softer if left to build up some mass on its own, and softer still if each spring it is cut to the ground and lightly raked to remove dead parts. The overall effect is of a rough meadow, grayish green in summer and beige-tan in winter. It is not recommended for small gardens, as it spreads aggressively, although this same character may recommend it for other situations. It does best in sun with some summer water and is tough once established. Propagation by division is easiest, but it can be grown from seed.

Elymus californicus, California bottlebrush grass

Prefers: low elevation, cool temperatures, coastal habitats, moisture, part shade
Accepts: full sun near coast with moisture
Blooms: early spring, summer

California bottlebrush grass is a bold, three- to six-foot-tall, somewhat coarse grass arising from a loose cluster of inch-wide, bristly, pale green, gracefully drooping leaves. The tall spikes are topped by drooping, bristly flowerheads that resemble a

Elymus californicus

flaccid bottlebrush. Native to but uncommon in moist canyons and coniferous forests of the coastal Bay Area, this grass is on the CNPS "watch" list (List 4). Seeds and plants are available from specialty nurseries and seed suppliers.

This cool-season grower is easily kept green with summer irrigation and light shade. It is dramatic when grown singly or in small colonies in woodland margins, as an accent among shrubs, at waterside, or in the back of a wet meadow. It prefers a cool coastal climate with some irrigation, but could be tried in hotter interior climates with some shade and regular irrigation. It looks best if renewed in late summer by cutting back hard. The bristly hairs on all parts of this plant can be irritating to the skin. It self-sows and is easy from seed or by division.

Elymus cinereus, see *Leymus cinereus*

Elymus condensatus, see *Leymus condensatus*

Elymus elymoides ssp. *elymoides*, squirreltail
Prefers: mid- to high elevation, sun, dry, good drainage
Accepts: poor soils, including serpentine, heat, drought
Blooms: late spring, early summer
Squirreltail is a widespread grass of poor soils in dry, open areas throughout much of the West. It is frequently overlooked because of its unremarkable appearance for much of the year, but it can be quite showy in flower and seed. The clump is usually less than a foot tall and wide, with narrow, grayish green leaves, and even at its best it would not be grown for its foliage.

In late spring a soft-looking, narrow plume, much like a long, thick artist's paintbrush, extends about eight inches above the foliage. The plume is silvery pink, almost iridescent, subtle yet showy. As the clump continues to dry out, this narrow brush suddenly opens out so that the long bristles of the plume point outward like the hair on a squirrel's tail. The expanded, bristly, ivory/straw-colored plume creates a display of great beauty, especially when backlit by the sun.

As the plume ripens, it soon breaks apart, and its individual components blow around like tiny tumbleweeds, as if searching for a suitable crevice to wait out the blazing summer until the next season's rain. Only the bottom-most cluster of bristles remains on the top of the stem. As attractive as the flowering display is in the wild, in the garden it seems rather brief, maybe a month or two at the most. The starlike bristles on the stalk persist into winter, while the foliage clump dries.

Despite its drawbacks, squirreltail has its place, especially in larger landscapes with areas of poor, stony soil. It is especially tolerant of serpentine. This grass accepts heat and drought, and it could be used in dry meadows or in desert gardens, where other plants would carry the display in its off-season. It is easy from seed and self-sows readily in open sites, but it does not compete well.

Elymus glaucus, blue wildrye
Prefers: mid-elevation, sun, dry, good drainage
Accepts: summer watering
Blooms: late spring, early summer
Blue wildrye is a modest and variable bunchgrass, valued primarily for its tough constitution, which makes it useful for revegetation and erosion control. The foliage is close to a foot tall with leaves about one-quarter inch wide. Two- to three-foot spikes emerge in late spring, usually strictly erect or slightly arching, with bristle-tipped flowers pressed upright against the stalk, producing a narrow inflorescence.

Blue wildrye self-sows readily, tending to germinate within the crowns of other plants, and its massive root system makes it difficult to remove. However, where its invasive character is not a concern, such as in large-scale, naturalistic plantings, it provides a strong vertical accent with foliage varying from bright, deep green through gray. A cool-season grower native to open areas, grassland, chaparral, and forest throughout California to Alaska, the Great Plains, and New Mexico, this grass is easy from seed or by division.

There has been an unfortunate confusion in the nursery trade where *Leymus mollis* (formerly *Elymus mollis*) has been called *Elymus glaucus*, probably because of the glaucous leaf color of *L. mollis*. The two are quite distinct, *E. glaucus* being a clump-forming grass with narrow leaves, while *L. mollis* travels extensively underground and has broad gray leaves over half an inch wide.

Elymus mollis, see *Leymus mollis*

Elymus multisetus is similar to *Elymus elymoides* ssp. *elymoides* and has similar uses.

Elymus triticoides, see *Leymus triticoides*

Festuca brachyphylla, alpine fescue
Prefers: high elevation, sun or part shade, regular water, excellent drainage

Blooms: summer

This tiny alpine fescue is only two to three inches across and forms dense tufts of gray or green, threadlike foliage. Flower spikes rise a few inches above the foliage. Native to gravelly ridges of the central and southern high Sierra Nevada and the White and Inyo Mountains, this is a charming grass for a regularly watered rock garden with loose, well-aerated soil. Propagated from seed or by division, it may increase slowly by self-sowing.

Festuca californica, California fescue

Prefers: low to mid-elevation, part shade, some moisture
Accepts: full sun with routine irrigation
Blooms: spring, summer

California fescue is a handsome and useful medium-sized bunchgrass that forms dense clumps a foot or so high and slightly wider. Foliage color varies from green or blue-green to gray-blue, and the leaves, which arch stiffly toward the ground, are twenty inches long and one-quarter inch wide. The three- to four-foot flower stalks emerge from the center of the clump and arch elegantly outward. The inflorescence is loose and open, with the flowers dangling from the arching tips of the

Festuca californica. (SAXON HOLT)

Festuca californica

plume's branches. Initially the stalk and inflorescence are similar in color to the foliage, but by flowering time, tints of gold and salmon shine through, ultimately aging to straw color.

California fescue is native to open, rocky, brushy or wooded slopes, especially north-facing banks in grassland, chaparral, and coniferous and deciduous woodland (particularly Oregon oak, *Quercus garryana*) in central and northern California and Oregon. This grass makes a wonderful specimen, but may look most natural when displayed in small groups or large drifts. It is tolerant of full sun if given occasional water, but is best with light shade, particularly if planted away from the coast. It is easy from seed or by division.

California fescue is best in central through northern California (excluding the hottest areas) as a large-scale groundcover on banks and steep north-facing slopes, where it provides a lovely textural effect, and it works well as an understory grass among taller, open-canopied shrubs or trees. It is also an excellent grass to bridge the transition between a meadow and shrub or wooded zone. When used in large swards, space some closely to approximate pure stands, and space others more distantly to permit other herbs, grasses, or bulbs to coexist. In southern California, or in hot interior valleys, thorough bi-weekly watering and some shade are essential, but even with water the longevity of the plant and ease of cultivation decline dramatically in these areas.

Although the following cultivars, to be true to name, must be from divisions, many produce seedlings similar to the named form, and an alert gardener may find some that are even better.

'**Blue Fountain**', with bright gray-blue foliage, is a selection made by Nevin Smith from Middle Blue Ridge in Henry Coe State Park in Santa Clara County. The clump is compact and produces prolific flower spikes to about four feet.

Two selections made by Sally Casey from Horse Mountain in Humboldt County are '**Horse Mountain Gray**', with wide, very gray leaves, and '**Horse Mountain Green**', with drooping leaves bright green above and dull gray-green beneath, giving a bi-colored effect.

'**Serpentine Blue**' is a very blue-gray selection by Roger Raiche from Pine Mountain in Marin County on serpentine soils. It seems to be particularly strong in holding its inflorescences into late summer, and also seems to be the most widely grown cultivar to date, performing well even in hotter climates at the University of California Arboretum at Davis and in southern California at Rancho Santa Ana Botanic Garden.

'**Mayacamas Blue**' is a vivid blue-gray selection made by Roger Raiche from the Mayacamas Mountain range north of Mount St. Helena in Lake County, where it formed an understory in ponderosa pine woodland.

'**Blue Select**' is a seed strain with blues selected.

'**Fatso**', introduced by Phil Van Soelen at California Flora Nursery, is a robust plant with wide green leaves.

Festuca idahoensis, Idaho fescue
Prefers: low to high elevation, good drainage, sun or part shade, some irrigation
Accepts: sun or shade
Blooms: early spring, summer
Idaho fescue is a beautiful and variable grass that forms dense tufts of basal leaves two to twelve inches long and often gray to gray-green, although green-leaved forms occur. Flowering stalks are one foot to three feet tall, topped by a loose but narrow panicle, giving an elegant fountainlike effect. Unfortunately, the taller scapes tend to break easily after drying. This cool-season grass is native to open woodlands, meadows, and rocky slopes from near sea level in the San Francisco Bay Area up into its predominant home in the mountains of central and northern California to Colorado and Canada.

There are forms suited to many garden types and climates. Lower-elevation plants, which can be found in a range of leaf colors, seem to do well in cooler areas of central and northern California in full sun to light shade, although they benefit from some (even regular) summer water. Once established, they can be grown dry, but may have a somewhat crisp appearance. They work nicely in masses to create a hummocky meadow, in drifts at the edges of woodlands, or as specimens in irrigated rock gardens. The dwarf alpine forms, often gray-blue, are more difficult to grow, need excellent drainage and regular water, and may be best left to rock gardeners or grass specialists. Plants are propagated from seed or by division.

Some garden-worthy cultivars of Idaho fescue include the following:

'**Tomales Bay**', selected by Walter Earle in Tomales from a seed batch collected by Judith Lowry near Nicasio Reservoir in Marin County. It is easy to grow, with gray-green leaves in six- to eight-inch clumps.

'**Eagle Peak**', with gray-blue foliage, is a selection by Nevin Smith from the Warner Mountains in Modoc County. It has compact rosettes six to eight inches tall with scapes about eighteen inches tall, sometimes showing a pinkish tinge at flowering.

'**Stony Creek**', from Del Norte County, was introduced by the East Bay Regional Parks Botanic Garden. It forms large, durable, blue-gray foliage clumps and serves well as a native alternative to the nonnative *Festuca ovina*.

'Muse Meadow' is a selection by Nevin Smith from the Marble Mountains near Boulder Peak. It is tightly clumping with darker, bluish green foliage.

'Snow Mountain' is a Nevin Smith selection from Snow Mountain at the nexus of Colusa, Lake, Glenn, and Mendocino Counties. It is larger-growing, forming clumps about ten inches across, and has blue-gray, disease-resistant foliage.

Festuca 'Siskiyou Blue', once thought to be an especially gray-leaved form of the California native *F. idahoensis*, is now believed to be a form of *F. ovina*, a European species.

Festuca occidentalis, western fescue
Prefers: low to mid-elevation, light shade, some moisture, good drainage
Blooms: early spring, summer
Western fescue makes fine, dense tufts that remain green year round with some summer water and a light mulch of litter such as short-needled pine. The four- to twelve-inch-long leaves are soft, smooth, and bright green. Delicate flower panicles on stems from two to three feet tall rise above the elegant clumps of this cool-season grass. Native to conifer, oak, and mixed evergreen woodlands, mostly in the low- to mid-elevation mountains of central California north to British Columbia, western fescue can be difficult in cultivation. Where successful, it provides a welcome and cheerful accent in areas with light shade. It is grown from seed or by division. A colorful contrast would be to combine it in drifts with a blue-gray selection of *Festuca idahoensis*.

Festuca rubra, red fescue
Prefers: low to high elevation, part shade, routine water
Blooms: early spring, summer
Red fescue has a worldwide distribution and is highly variable even within California. It is a fine-textured, slowly spreading grass with thin leaves six to twelve inches long and flower stalks sixteen to forty inches tall. It forms thick, irregular tufts, creating a tussocky texture, and can be cut for more even turf. Some forms of this species are commonly used in turf mixtures, although no California natives have yet been used this way. This cool-season grass is found on sand dunes, in grassland, and in other open, moist places from sea level into the high mountains, and it requires regular moisture for best garden appearance. A patch of this grass at the height of bloom is memorable, bright green to almost metallic blue-gray and glinting in the sun.

In California there are two color forms. Bright deep green is most typical of the species, but in coastal areas, especially along the north coast, there are some lovely, low-growing, blue-gray forms. In cooler climates of the state, it is excellent as a moist meadow grass and blends well with moist-growing bulbs and perennials, as well as with other moisture-loving grasses. Annual mowing and raking out of dead thatch prior to the fall growing season helps to maintain a healthy, thick stand. In southern or hotter regions it is best as an informal accent; it is not successful as a solid mass planting. In these regions it declines in appearance during the summer months. It is grown from seed or by division.

The cultivar **'Molate Blue'**, selected by David Amme from Point Molate in Contra Costa County, has greenish gray foliage in dense tufts and a tall, powdery gray inflorescence scape. Summer dryness accentuates its gray color, and it is much more tolerant of drought and sun than other selections.

'Jughandle' is another selection by David Amme from Jughandle State Park on the Mendocino County coast. It has short, dense, blue-gray foliage three to five inches tall, with flower scapes about eight inches tall. It does well in sun, but is not tolerant of drought or heat.

'Patrick's Point' is a selection by Roger Raiche from Wedding Rock at Patrick's Point on the Humboldt County coast. It is vivid blue-gray, short and dense, with short flower spikes that are barely noticeable. It would make an interesting blue-gray lawn or meadow. As a mixed border accent, it provides an exciting counterpoint to bronze- or chartreuse-foliaged plants such as the bronze-leaved stream orchid, *Epipactis* 'Serpentine Night'. Unfortunately, there has been a mix-up by nurseries, and two different plants are sold under this name. The true 'Patrick's Point' spreads underground, while the impostor is tightly clumping.

'Jana's Blue' is a selection by Jana Niernberger from near the lighthouse at Point Arena on the Mendocino County coast. Its foliage is blue-gray, but with more green than in 'Jughandle' and 'Patrick's Point', and both foliage and scape are somewhat taller than those of 'Jughandle'.

Hesperostipa comata, needle-and-thread
Prefers: mid- to high elevation, heat, dry, full sun, rocky to sandy soils
Accepts: drought, some summer irrigation
Blooms: spring, early summer
This cool-season bunchgrass is striking in flower. From its clump of narrow, grayish green basal leaves, strong one-foot to three-foot scapes shoot out, topped by four to twelve inches of narrow, spikelike panicles from which project astonishingly long

Hesperostipa comata. (STEPHEN INGRAM)

(four to eight inches), flexible, and irregularly twisted awns on the ends of the seeds. The effect is lovely, with the long awns twisting sensuously in the breeze.

Native to the high mountains of the eastern Sierra Nevada, Great Basin, and southern California deserts, as well as Arizona, Idaho, Montana, Utah, and Colorado, this grass has proven temperamental at lower elevations. Best in well-drained soils in full sun and heat, it tolerates occasional summer water, but the tall scapes do not hold up well when wet. It provides a strong vertical accent behind fine-textured grasses, and the scapes are attractive in floral arrangements. It is grown from seed or by division.

Hierochloe occidentalis, sweetgrass, vanilla grass

Prefers: low elevation, good drainage, rich soils, some moisture, part to full shade
Accepts: heavy clay soils, morning sun if irrigated
Blooms: early spring, summer
Sweetgrass is a delightful perennial bunchgrass with bright green, ribbonlike leaves, blue-gray on the undersides and fragrant when stroked, crushed, or dried. Dainty white flowers on two- to three-foot stalks bloom from late winter through summer on this cool-season grower. Sweetgrass is native to redwood groves and forested canyons in humus-rich soils and deep shade or part shade in the North and Central Coast Ranges and the San Francisco

Hierochloe occidentalis. (ROBERT CASE)

Hierochloe occidentalis

Bay Area to Washington, where it thrives without summer water. It needs some irrigation and part to full shade in most California gardens.

Although tolerant of heavy clay, sweetgrass does best in rich soils. An excellent grass for shady woodlands and north-facing slopes, the foliage retains its deep green color throughout the year and combines beautifully with ferns and other woodland herbs such as redwood sorrel (*Oxalis oregana*) and deerfoot (*Achlys* spp.) or *Vancouveria hexandra*. It can be grown from seed or, more easily, by division. It is uncommon in the nursery trade because it seems to dislike container culture.

Hilaria jamesii, see *Pleuraphis jamesii*

Hilaria rigida, see *Pleuraphis rigida*

Hordeum brachyantherum, meadow barley
Prefers: low to high elevation, sun, vernally wet soils
Accepts: clay soils
Blooms: spring, summer
Meadow barley is a short-lived perennial grass forming loose clumps of flat blades less than half an inch wide and six inches long. This cool-season grower is topped in spring by narrow spikes of densely packed, upward-facing bristles that are tinged purple, like a small version of the classic image of a cereal-grain seedhead.

Native to a wide range of open, vernally wet meadows and streambanks in most parts of California (except the deserts) to Alaska, Mexico, and far beyond, meadow barley blends nicely in a mix of native meadow grasses. In cultivation the individual clumps may seem somewhat shabby, although it could be massed as a groundcover in soils wet at least in spring, where the numbers would compensate for its solitary shortcomings and the hundreds of flower spikes would create a pleasing effect. Meadow barley is easy from seed or by division.

Hystrix californica, see *Elymus californicus*

Koeleria macrantha, Junegrass
Prefers: low to mid-elevation, sun, dry, good drainage
Accepts: some irrigation, part shade
Blooms: spring, summer
Junegrass is a good-looking perennial grass forming a short, round tuft of narrow leaves six to twelve inches long and one-foot to two-foot flower stalks with stout,

spikelike panicles of pale green florets that stand out in a mix of grasses with more lax and open panicles. The flower panicle starts out narrow, opens wide in bloom, and then may return to its narrow shape.

This cool-season grower is native to dry, open sites in clay to rocky soils, shrublands, woodlands, and coniferous forests in many areas of California to Alaska, Canada, central and eastern United States, and northern Mexico. It is attractive in a moist or dry meadow mix of low grasses and associates. In moist locations or in stagnant air, rust fungus may attack the foliage, but this can be easily clipped off and the plant renewed. Junegrass is grown easily from seed or by division.

Leymus cinereus, ashy wildrye, Great Basin wildrye
Prefers: mid-elevation, sun, heat, dry, good drainage
Accepts: some summer irrigation in hot areas
Blooms: early spring, summer
This is a large-scale, bold-foliaged bunchgrass with leafy stems three to five feet tall in a dense clump and vigorous pale green, gray-green, or gray-blue leaves that dry to tawny straw color in summer. The thick, bristly flower spikes, about six inches long, add strong vertical character to the clump.

Native to streamsides, canyons, sagebrush flats, and open woodlands in the Cascades, eastern Sierra Nevada, Great Basin, and eastern desert mountains of California to Canada and Colorado, this grass makes a bold, eye-catching clump in dry landscapes. A cool-season grower in mild climates and a warm-season grower in areas with cold winters, it does best in hotter regions of California, sulking and developing rust fungus in cool coastal gardens.

Ashy wildrye provides a dramatic accent in large landscapes, and can be massed for screening, to highlight architectural schemes, as a complement to large shrubs, or to add height or focal points in a meadow garden. It is best with monthly watering in summer, and it should be cut to the ground in winter. It is grown from seed or, less easily, by division.

Leymus condensatus, giant wildrye
Prefers: low to mid-elevation, sun, some moisture, good drainage
Accepts: drought, part shade
Blooms: spring, summer
Giant wildrye is one of the largest California grasses, growing to twelve feet tall. This somewhat rangy plant is native to dry slopes and open woodland in south-

western California, the Mojave Desert, and Mexico. The flat, rough leaf blades, to one inch wide, occur mostly on tall stems and remain green most of the year. Typical leaf color is bright green or deep green, but plants from the South Coast Ranges and southern California can be gray or blue-gray. The dense flower spikes are six to eighteen inches long and showy, but often are not produced in cultivation.

In the garden giant wildrye needs occasional watering through the dry season to look its best, and it is quite tolerant of frequent watering. It spreads by short, heavy rhizomes, but spread is typically less than one foot a year. A dramatic plant that needs room to sprawl, it tends to lean on adjacent shrubs or trees. It can be effective as screening or background in larger gardens.

For a more compact plant, tall shoots may be pruned back and foliage cut down in winter, which will reduce the plant's vigor and density. The gray-foliaged types seem happiest in full or nearly full sun. These can provide dramatic accent to meadows or mixed shrub plantings. Giant wildrye is easy from seed or by division.

The cultivar **'Canyon Prince'**, introduced by the Santa Barbara Botanic Garden, is from Prince Island, near San Miguel Island in the Santa Barbara Channel. It is best in sun, where it is shorter and more erect than the species. Spreading underground even more readily than the species, it is recommended only for larger gardens or confined to containers. The blue-gray leaves are perhaps the most stunning of California's larger native grasses. The flower spikes, on three- to five-foot stems, are produced prolifically in some situations and rarely in others. This selection brings dramatic color and form to the garden, and it seems to thrive wherever it has been tried in California. Some experts believe that 'Canyon Prince' may be more closely related to *Leymus mollis* than to *L. condensatus*, and this question deserves more research.

Leymus mollis, dune ryegrass

Prefers: low elevation, sun, some moisture, good drainage
Blooms: summer

Dune ryegrass is a rapidly spreading grass native to beaches and sand dunes from central California northward around the Pacific into Asia. It has broad, half-inch- to one-inch-wide, flat, gray or blue-gray leaf blades on two- to five-foot stems and dense flower spikes four to twelve inches long. Although the foliage is bold and the color is good, the stems tend to flop, the clumps are insubstantial, the foliage goes dormant in winter, and the underground spread, sometimes more than ten feet in a single season, makes it unmanageable for all but the largest gardens. It can be confined to a container if the container is not resting on the ground.

This grass is useful in its native habitats for dune stabilization and erosion control, or as a block of colored foliage in an enormous meadow planting, but in almost all regards it would be less trouble to use the more attractive and manageable *Leymus* cultivar 'Canyon Prince' (see *L. condensatus,* above). Dune ryegrass is easy from seed or by division. This plant sometimes has been sold under the name *Elymus glaucus*, which is a very different grass.

Leymus triticoides, creeping wildrye
Prefers: mid-elevation, sun, some moisture
Accepts: heavy soils, salinity
Blooms: late spring, summer
This is a rapidly spreading, even invasive, grass that produces straight stems two to four feet tall, with slender, stiff leaves and narrow, erect spikes to eight inches that eventually nod. Color varies enormously in the wild, from blue-gray to deep green. It spreads underground by rhizomes (to ten feet in a season) and retains its color into the summer when many other grasses have gone dormant.

Native to moist, often saline, mountain meadows throughout much of California (except the deserts), north to Washington and south to Texas, creeping wildrye grows on heavy soils and is valuable in preventing erosion, especially along watercourses. It thrives even in ungrazed grassland, competing well with exotic grasses, where smaller native grasses often disappear. Because of its invasiveness, it is recommended only for large landscapes or container culture. In irrigated landscapes rust fungus can be unsightly. It is usually grown by division, as seed is seldom produced.

'**Gray Dawn**' (sometimes sold as "Gray Ghost"), to two feet tall, a selection by Nevin Smith from Watsonville, has blue-gray foliage with an almost metallic sheen.

'**Shell Creek**', a selection by David Fross from near Shell Creek in eastern San Luis Obispo County, is blue-gray, dense-growing (though still spreading), and only about eighteen inches tall.

Melica californica, California melic
Prefers: low to mid-elevation, sun, dry, good drainage
Accepts: part shade, most soils
Blooms: early spring, summer
California melic is sparkling in flower, with sometimes purple-tinted, translucent, papery bracts on narrow, twelve-inch spikes atop two- to four-foot, straight stems

and one-foot-tall clumps of flat basal foliage. Native to open meadows, hillsides, and oak and conifer woodlands of central and northern California and the Sierra Nevada foothills, this cool-season grower is tolerant of many soils.

Best grown dry in full sun or light shade, California melic is dramatic as a specimen or in small groups on slopes, in meadows, or at the margins of or in openings within shrubs. Cut flower spikes are attractive in floral arrangements. It is easy from seed or by division.

'**Temblor**', a selection by David Fross from the Temblor Range in eastern San Luis Obispo County, forms tight foliage clumps and has exceptionally fat flower spikes.

Melica frutescens, reed melic
Prefers: low to mid-elevation, sun, dry
Accepts: light shade
Blooms: spring, summer

Reed melic is a little-known melic from dry slopes, chaparral, and woodland in the inner South Coast, Transverse, and Peninsular Ranges and deserts of southern California to Arizona and Baja California. Strong stems from two to six feet tall are topped by four- to sixteen-inch-long spikes with purple-tinged, translucent papery bracts.

This grass is an excellent accent for sun or light shade in a dry meadow, shrub planting, or thin woodland. Seldom grown, it deserves to be tried more widely in gardens. Plants are grown from seed or by division.

The cultivar '**Palomar**', a selection by Mike Curto from Mount Palomar, has foliage clumps twelve to eighteen inches tall and flower spikes on stems to three feet.

Melica harfordii, Harford's melic
Prefers: low to high elevation, light shade, cool, some moisture
Blooms: spring, summer

This melic forms tall, almost columnar, loose clumps about one foot wide, with stems two to four feet tall, the top four to ten inches consisting of a narrow spike lined with pale papery bracts that catch the light exquisitely. A lovely grass from dry slopes, canyons, and coniferous forests from central California north to British Columbia, it is best grown in cooler northern California gardens and benefits from watering every few weeks in the dry season.

Harford's melic adds verticality and delicacy to lightly shaded slopes, meadows, and open shrub or woodland plantings, either as an accent or in mass. Dried inflorescences are attractive in flower arrangements. It self-sows freely on favorable sites and is easy from seed.

Melica imperfecta, melic

Prefers: low to mid-elevation, full sun to part shade, some moisture
Blooms: early spring, summer

Melic is a variable species of delicate appearance that grows on dry, rocky hillsides, in chaparral, or in open woodland from the Santa Cruz Mountains, the central and southern Sierra Nevada, and the South Coast Ranges to the western Mojave Desert and Baja California.

A cool-season grower, melic forms loose to dense clumps under a foot wide with narrow, usually glossy, nearly evergreen leaves and many wiry flower spikes from eighteen to forty inches tall. The inflorescence may look like a narrow spike, or the side branches, which are strictly held against the stem in the "spikelike" types, can arch upward, outward, or even downward in substantial branched plumes, with tiny bracted flowers at the tips.

A subtle yet satisfying addition to meadow, open shrub, or sparse woodland plantings, melic does best in full sun in cooler northern climates and in part shade in southern or hotter areas. It is particularly effective with woody subshrubs such as *Monardella* or *Eriogonum* species, growing out of mat-forming groundcovers, or softening a hard-edged succulent or desert scheme. It self-sows in open soils and is easy from seed or by division.

Melica torreyana, Torrey's melic

Prefers: low to mid-elevation, good drainage, dry, part shade
Accepts: sun near coast, some summer irrigation
Blooms: early spring, summer

Torrey's melic forms clumps of soft, flat, narrow leaves with thin, one-foot to three-foot stems, erect in the cen-

Melica torreyana. (WILLIAM FOLLETTE)

ter of the clump and arching out and sideways around the margin. Its thin flowering spikes appear as if grains of white rice are sprinkled evenly along its length;
these are the translucent papery bracts characteristic of most plants of the genus.

Torrey's melic is like a more refined version of *Melica californica*, and it shares

Melica torreyana

much of that plant's tough constitution and horticultural uses, although it probably does best with more shade. Native to chaparral and coniferous forests in the hills, canyons, and mountains of central and northern California, this cool-season grower should be cut back hard when dormant to encourage good form. The inflorescences make pleasing accents in flower arrangements.

Torrey's melic grows easily from seed or by division. Some plants from exposed, windy sites have flowering stems that are low to completely prostrate, although none of these has been named or is currently offered in the nursery trade.

Muhlenbergia montana, mountain muhly

Prefers: mid- to high elevation, sun, some moisture, excellent drainage
Accepts: low elevation
Blooms: summer

Mountain muhly is a diminutive mat-forming grass, producing low, spreading tufts about four inches tall of densely packed grayish leaves, often tan at the tips. The tiny flower scape is six to eight inches tall. Native to high elevations from the Klamath Ranges through the Sierra Nevada south to Mexico and Guatemala, it is found on rocky soils on open slopes and in dry meadows.

This grass requires very sharp drainage and watering every week or so. An attractive accent with other diminutive alpines, in a miniature meadow, or with watered succulents, it deserves to be tried under wider cultural conditions. It is grown from seed or by division.

Muhlenbergia richardsonis, mat muhly

Prefers: mid- to high elevation, sun, some moisture, excellent drainage
Accepts: low elevation
Blooms: summer

Mat muhly is similar to *Muhlenbergia montana*, but is more spreading, forming irregular green mats only three to four inches tall and three or more feet across. The inflorescence, a narrow spike, is four to six inches tall. Native to mid- to high-elevation moist meadows, streamsides, or talus slopes throughout much of California to Canada, the northeastern United States, and Mexico, it is a good rock garden subject. More garden experimentation is needed before recommending this plant generally, but it is a fine choice for the grass enthusiast.

Muhlenbergia rigens, deergrass

Prefers: low to mid-elevation, full sun, some moisture

Muhlenbergia rigens. (SAXON HOLT)

Accepts: drought when established
Blooms: late spring, summer
Deergrass is a vigorous and low-maintenance bunchgrass that makes dense, leafy domes two to three feet tall by four to five feet wide composed of thousands of thin leaves, with dozens of erect and arching, rodlike flower stalks that extend two feet beyond the foliage. The entire plant, fine-textured because of the thin leaves and scapes yet substantial in form, yields a fascinating mix of contrasting effects. The subtle blend of pale green, tan, beige, and gray in the leaves and gray in the scapes gives the plant an overall grayish brown cast that catches the light nicely. When left on the plant, the stalks remain attractive for twelve to eighteen months. Color does not change appreciably with the seasons, although summer growth has a freshness not seen in winter. New growth is fully green after plants are cut back, which should be done every three or four years.

A grass that seems to combine both warm- and cool-season growth patterns, deergrass is at home in full sun and prefers water but tolerates drought once established. Its extremely tough root system makes this a permanent plant in the landscape. It is native to sandy or gravelly streambanks and stream bottoms, seepages, and even dry canyons throughout the lower mountains and valleys of much of California south of Shasta County to the Mojave Desert, Texas, and Mexico. Flower stems of this grass were the primary foundation material for coiled baskets made by many Native American tribes from central and southern California.

Deergrass can be used in almost any sunny landscape, but it loses vigor and form in shade. It tolerates almost any climate, soil, and watering regime, and it is incomparable at filling open space with a tough, good-looking plant that suppresses weeds and holds the soil. In smaller landscapes it is a valuable accent with form and color unique in the mid-sized grasses, and it is suitable as a hedge, background, or transition to a larger landscape element. It is particularly effective paired with spike dropseed (*Sporobolus airoides*), which is close in scale and color but has a very different inflorescence.

Stalks of deergrass are used in dried arrangements. It is easily grown from seed, but propagation by division of mature plants is difficult. This grass varies in the wild, some forms being tightly clumping, others layering out at the margin of the clump. Some forms, especially those from higher elevations, are smaller and more compact in all features. However, none of these variants is yet named, and the standard nursery offering is the large, tightly clumping form.

Nassella cernua, nodding needlegrass
Prefers: low to mid-elevation, sun, dry, good drainage
Blooms: spring, early summer

Nodding needlegrass is perhaps the most elegant of California's native needlegrasses. It forms fine-foliaged clumps under a foot tall, from which two- to three-foot erect spikes emerge in spring. The flowering spike arches over at the tip and branches into a loose, dangling cluster of flowers, each producing a seed that is tipped by a long, irregularly bent, flexuous bristle or awn. In spring and early summer the pendulous clusters of bracts and bristles sway languidly with the slightest breeze or whip about wildly in the wind. Backlit by the sun, they recall the splendor of California's pre-European grasslands with an almost melancholy beauty. As with other *Nassella* species, the plants gradually bake in summer to a dried straw color, with only a remnant of green in the center of the foliage clump. This is a good time to cut or mow the plants and remove debris.

Native to grassland, chaparral, and juniper woodland in more arid areas of the North and South Coast Ranges and the Sacramento Valley through southern California and into Baja California, this is a grass of remarkable beauty for those willing to accept summer dormancy. It is dramatic when displayed on a wall or in other elevated situations, and it is spectacular massed in a meadow, provided drainage is good. It self-sows readily in open soil and is easy from seed or by division.

Nassella lepida, foothill needlegrass
Prefers: mid-elevation, dry, sun or part shade, good drainage
Accepts: some summer irrigation
Blooms: early spring, summer

Foothill needlegrass is like a smaller, more delicate version of *Nassella pulchra* or a less flamboyant version of *N. cernua*. It forms fine-foliaged clumps about eight to twelve inches across with many graceful stems two or more feet tall topped by loose, airy panicles with nodding branches. The bent awns that project from the flowers are smaller and shorter than those of either *N. cernua* or *N. pulchra*, but still are effective at catching the light and accentuating the motion of wind.

Native to dry slopes, chaparral, and oak woodland through the Coast Ranges and northern Sierra Nevada foothills to Baja California, this cool-season grower is useful in a variety of garden situations. It is one of the best grasses for meadows, and it fills empty spaces by self-sowing, making a solid clumping turf that will accept occasional mowing. It is also a good filler among shrubs or in open woodland, tolerating dryness, light shade, and many soils. Dry-looking after June when it enters dormancy, foothill needlegrass will green up again if given supplemental water. It is easy from seed or by division.

Nassella pulchra, purple needlegrass
Prefers: low to mid-elevation, sun, dry, good drainage
Accepts: most soils, high shade, some summer irrigation
Blooms: early spring, summer

To many enthusiasts of the native flora, purple needlegrass symbolizes pre-European grassland in California, of which it was a major component. Its persistence today is seen as a triumph over the massive environmental changes and loss of

Nassella pulchra. (WILLIAM FOLLETTE)

habitat that have occurred in the last two centuries. Some of the better surviving prairies of purple needlegrass, along with Idaho fescue and other native grasses, provide a fragmentary glimpse of a largely extirpated California landscape.

This cool-season grass is native to oak woodland, chaparral, and grassland in the valleys and foothills of northwestern California, through the Central Valley, and along the central and south coasts, on the Channel Islands, and into Baja California. It forms attractive bunches of fine, flat blades six to twelve inches long and showy panicles with tailed seeds formed by a bent bristle or awn. The open panicle branches often nod at the tips, and the long awn moves with the slightest breeze.

This stately and expressive grass is one of the best for meadow plantings and

Nassella pulchra

for controlling erosion on open slopes, and it will tolerate light shade and a diversity of soils. It is a dry grower that can be kept green longer with some summer water. In hot areas infrequent watering may be best, especially on heavy soils, although on well-drained soils it will accept more watering. When dormant it can be mowed to the ground to encourage new growth and thicker form, or it can be left as a dry meadow.

Purple needlegrass shows considerable promise as a low-maintenance groundcover, but regionally specific management techniques still need to be developed. Seeds can be sown directly on prepared soil, but they must be pressed into the soil for germination. Plants self-sow, forming thicker colonies each year.

Oryzopsis kingii, see *Ptilagrostis kingii*

Panicum acuminatum, panicgrass
Prefers: mid-elevation, sun, moist places
Blooms: spring, early summer
Panicgrass is a leafy, moisture-loving grass native to marshes and streambanks over a large area of North America and parts of northern South America. Its leaves are one inch to six inches long by one-half inch wide and in winter are clustered in attractive, dense rosettes, with most leaves flattened near the ground. In spring leafy stems can grow up to three feet tall, though usually they are less than one foot; outward-spreading, they are topped by airy clusters of flowers. The hazy plume is about four inches long and looks like finely branched threads tipped with tiny beads.

This plant shares features of cool-season and warm-season grasses, and water will keep it growing into fall, although it tends to look rather shabby. Uncommon in the wild, it may become quite weedy in an irrigated landscape. Plants look best if the mass of untidy stems and dried inflorescences are removed in late summer, and look freshest when confined to a moist meadow or pond margin. Panicgrass self-sows and is easy from seed or by division.

Phragmites australis, common reed
Prefers: mid-elevation, sun, heat, moisture
Blooms: mid- to late summer
Common reed is a woody, bamboolike grass that can grow up to fifteen feet tall but usually is much shorter. This plant traditionally was harvested for roofing. The stiff, reedy stems are leafy—each glabrous, green or glaucous leaf eight to eighteen inches long and one-half inch to two inches wide—topped by a six- to twenty-

inch purplish to white plume. It is usually available in a gold-variegated cultivar called **'Variegata'**, which probably is a variant of the European form.

In California, common reed is found along lakes, rivers, and seepages, in marshes, and in sloughs. With an astounding worldwide distribution, it is perhaps the most widely distributed of all vascular seed plants. This is a warm-season grass that needs plenty of water and thrives in interior heat; it does not seem to do well in cool coastal climates. It spreads by rhizomes and can be invasive, but if restrained can serve as a dramatic feature of a large landscape. Stems and plumes are ornamental when dried.

Common reed is grown from seed or by division. It should not be confused with *Arundo donax,* also called reed, which is a stouter, taller, nonnative plant that is extremely invasive.

Pleuraphis jamesii, galleta
Prefers: low to mid-elevation, full sun, some moisture, good drainage
Accepts: drought
Blooms: mid- to late summer
Galleta is an attractive warm-season grass native to dry, sandy to rocky slopes, flats, scrub, and woodland in arid climates east of the Sierra Nevada and in the northern and eastern desert mountains to Wyoming and Texas. It forms a twelve- to eighteen-inch-tall, erect, spreading mound of pale green, leafy stems topped by grayish tan spikes about two feet tall. The flower spikes suggest fine beadwork.

Galleta needs full sun and good drainage and, although tolerant of drought, looks fresher when given some water during the dry season. The erect, almost reedy character of this grass makes a nice complement to desert shrubs, such as Mormon tea (*Ephedra* spp.), and its tough constitution and spreading habit suggest its use as a coarse mowed turf substitute. Dormant in winter, when it turns beige, it revives with the longer days and greater heat of early summer.

Little known in California gardens, galleta deserves greater experimentation. It is easy from seed or by division.

Pleuraphis rigida, big galleta
Prefers: low to mid-elevation, dry, full sun, good drainage
Blooms: summer
Big galleta is much like *Pleuraphis jamesii* but taller (to three feet), grayer, and finely white-woolly in all parts, with a coarser, woody-stemmed, somewhat sprawling growth habit. It is common on dry, open, sandy to rocky slopes and in flats, washes, dunes, scrub, and woodland in the deserts of southeastern California to

Utah and northern Mexico. This grass probably is best for large, arid landscapes combined with shrubs or to give a shrublike effect in a meadow. It is grown from seed or by division.

Poa secunda ssp. *secunda*, pine bluegrass
Prefers: low to high elevation, dry, good drainage
Accepts: poor soils
Blooms: spring, early summer

Pine bluegrass is a subtle yet lovely species that adds character and beauty to many natural areas in California from sea level into the high mountains. In early spring it produces a small tuft of foliage, usually less than six inches tall, from which thin, one-foot to three-foot spikes emerge, each topped by a soft inflorescence that has tints of pale green and purple when fresh. At low elevations the foliage often goes dormant at or near flowering time, and as the spikes dry they develop tints of salmon and gold.

Turning brown early, pine bluegrass is one of the first grasses to announce the impending dry season. It is native to many habitats in California, typically on stony or poor soils, and much of western North America to the Rockies, Alaska, and into South America. A cool-season grower at low elevations, it is more of a warm-season grass in higher-elevation meadows. Many ecological forms occur and intergrade. This is a lovely grass to add textural interest to a dry meadow or to openings in woodland or chaparral, although it seems to be short-lived in gardens. It is easy from seed.

Ptilagrostis kingii, King's ricegrass
Prefers: high elevation, sun, regular water, good drainage
Blooms: summer

King's ricegrass is a beautiful, fine-textured grass from alpine and subalpine meadows and streambanks in the Sierra Nevada. It produces golden green, hairlike foliage in small clumps four to six inches tall. Thin, delicate flower spikes rise twelve to eighteen inches above the foliage in summer. This grass needs extra care to perform well at lower elevations, preferring a cool, north-facing bank in well-drained soils with routine irrigation. It is lovely as an accent in a rock garden or container planting. It is grown from seed or by division.

Sporobolus airoides, alkali sacaton
Prefers: low to mid-elevation, sun, vernally moist conditions

Accepts: alkaline and clay soils, drought

Blooms: early summer

Alkali sacaton is a large, full, warm-season bunchgrass that combines toughness with surprisingly delicate beauty. The pale gray-green foliage forms round, two- to three-foot-tall clumps, the blades flat when watered and curled in on themselves when grown dry. Large, purplish, conical plumes with a stylized structure, much like a scale model of the branching form of a Douglas-fir, terminate two- to four-foot, sometimes longer, stems in summer and lend a wonderful delicacy to the plant.

Native to vernally moist valley bottoms or alkali sinks in central and southern California (including the deserts) to the central and southern United States and Mexico, alkali sacaton deserves wider use in gardens. The golden tan of its dried foliage during winter dormancy gives this grass more seasonality than its larger cousin, *Muhlenbergia rigens*. The two can be used in many of the same ways and are excellent in combination.

Tolerant of drought once established, and of alkaline or even seasonally submerged clays, alkali sacaton looks freshest and most ornamental in normal soils with occasional or even regular irrigation. The plant is much enhanced by positioning where the sun shines through the flower spikes. The ornamental spikes can be picked fresh or dried for arrangements. Plants are easily grown from seed, but are difficult to divide.

Stipa cernua, see *Nassella cernua*

Stipa comata, see *Hesperostipa comata*

Stipa coronata **var. *depauperata***, see *Achnatherum parishii*

Stipa lepida, see *Nassella lepida*

Stipa pulchra, see *Nassella pulchra*

Stipa speciosa, see *Achnatherum speciosum*

Stipa thurberiana, see *Achnatherum thurberianum*

SEDGES

The sedges represent an enormous group of plants in the Cyperaceae, a family of more than 100 genera and 3,000 to 5,000 species worldwide.[3] In California there are fourteen genera and over 200 species, but only plants of the genera of *Carex*, *Eleocharis*, *Eriophorum*, *Isolepis*, *Schoenoplectus,* and *Scirpus* are included here. The sedges are often described as both rushlike and grasslike. Those that are mainly stems with inconspicuous leaves, such as tules (*Scirpus* spp.) and spikerush (*Eleocharis* spp.), resemble some *Juncus* species of the rush family in their pattern of stems and leaves. Plants of the genus *Carex* have grasslike leaves—long, narrow blades with pointed ends.

Much like the grasses, sedges occur in many habitats throughout California, from coastal beaches to treeline in the mountains. In general, however, sedges are more moisture-loving than grasses, frequently are more evergreen, and thus are less seasonal in appearance.

Sedges were used by Native Americans for food, medicine, clothing, shelter, and other purposes. New shoots, pollen, seeds, or bulbs of some sedges were eaten or used medicinally, and leaves were used as a poultice on wounds or burns. More typically, utilitarian objects such as canoes, boats, rafts, mats, hats, skirts, sandals, water bottles, decoys, cages and corrals, and thatch and siding for dwellings were created from various sedges. Particularly good sites for harvesting sedges were highly valued and carefully managed.

Growing Sedges

In the garden sedges require more expertise and care than most native grasses. Many sedges have rapidly spreading rhizomes and are excellent soil binders, especially in moist areas. This character can be a liability in a small garden, as they can be difficult to control when planted in the ground. Containers provide a good solution, as most sedges grow well when confined. Another solution is to install continuous underground barriers, such as are used to control spreading bamboos.

Most sedges will grow whenever temperatures are regularly above freezing and water is dependable. In mild coastal climates, many will put out new leaves throughout the winter, although flowering is more typical in spring or summer. Many sedges, especially drier, woodland species, may go into partial summer dormancy if water is not provided, but they typically retain some green foliage. Supplemental irrigation will revive and prolong their growth. Most sedges will stop growing when winter temperatures routinely fall to near freezing. Some species, such as *Carex nudata*, may lose their leaves, while others may retain evergreen foliage until spring.

Sedges are not difficult to grow, and all can be started from seeds or rhizomes. Seeds are best planted in fall or early spring in pots or flats; they should be lightly covered with fine soil and kept moist until germination. When they have five or more leaves they can be moved to larger pots. Plants can be divided at any time of year if the divisions are kept moist. During the rainy season, sedge divisions usually can be planted directly in the ground, provided they do not dry out before new roots are established. Although the rapidly spreading sedges divide most easily, even tightly clumping ones will readily reestablish. Many of these plants are quite invasive by seed, and their extensive root systems can make them difficult to eradicate.

Once established, even moisture-loving sedges can tolerate some drought, but irrigation will keep them looking their best. Fertilizer, although usually not necessary, will make most actively growing sedges appear lusher and promote faster growth and spread. This may be desirable in garden situations where a verdant effect is desired or when a groundcover is being established.

Sedges are maintained much like grasses, with the notable exception that they do not regenerate quickly after heavy pruning or mowing. When a sedge is cut to the ground, which may be done every year or two to remove dead material, the plant often takes several months to come back to an approximation of its former self. Sedges that spread by underground runners revive more rapidly after cutting than those that form tight clumps, but it is best to wait until plants start their seasonal growth.

Another option is to dig up the old clump or colony and replant vigorous side shoots, which will grow actively during the coming season. However, given the tough root systems of most sedges, this is more work than most gardeners will be willing to undertake. For sedges grown in containers, emptying the pot, reconditioning the soil, and replanting with offsets may be the best approach, as the plants will look better than if they are simply cut back and left rootbound.

Sedges are used in the garden in much the same manner as grasses. Being generally more moisture-loving, they tend to fit best in irrigated landscapes such as rock gardens (for the smaller species), in moist meadows, or in borders mixed in with perennials or shrubs. Some of the larger sedges are effective only in large landscapes, and associated with large landscape features such as ponds, lakes, or streamsides.

Sedges with a spreading habit are ideal as soil binders along watercourses or wet banks, and some of the shorter species are being tried as low-maintenance turf substitutes. The drier sedges are excellent as understory elements in chaparral or woodland gardens or scattered in dry meadow plantings. Noticing how sedges grow in nature can give valuable insights about their uses in the landscape. However, even most of the clumping, nonrhizomatous species are invasive by seed, and their extensive root systems can make them difficult to eradicate. Plant combinations should be chosen with care.

Sedges for the Garden

Several hundred sedges are native to California, mostly in the genus *Carex*, but there is little first-hand experience in growing this enormous group of plants. Also, sedges are difficult to "key out" to their correct botanic names, and many people do not know the botanic names of the sedges they grow, so it is difficult to document experience with particular species.

Carex amplifolia, bigleaf sedge
Prefers: low to high elevation, sun or shade, wet places
Accepts: some dryness in shade
Blooms: spring
Bigleaf sedge is a bold, attractive sedge with raised ribs on either side of a deep central trough on the two- to four-inch-wide, blue-gray to lime-green leaves. The leaves are three feet long with nodding tips, and the flowers are on stems to four feet tall, with nodding strings of greenish tan seed clusters.

This sedge is at home in wet marshes and bogs from central and northern California to Idaho and British Columbia. It spreads by underground runners and can quickly form a sizable colony in wet soils, which makes it ideal for naturalizing around large ponds or streamsides in full sun or shade but potentially overwhelming in small gardens. If grown with only occasional water, spread is much slower and the plant will be smaller in all aspects. In these circumstances, it will benefit from shade to prevent burning of the foliage. For a similar but nonrunning sedge, see *Carex spissa*.

Carex barbarae, Santa Barbara sedge
Prefers: low to high elevation, sun or part shade, some moisture
Blooms: spring
This is a large, invasive sedge about one to four feet tall with troughed leaves about an inch wide and taller, nodding inflorescences with dark brown seeds. It

occurs in large colonies in wetlands or on vernally moist, brushy or open slopes in southern California, the San Francisco Bay Area, the Cascade and North Coast Ranges, the Sierra Nevada foothills, and the Klamath Ranges into Oregon. Roots of this sedge were used by Native Americans for basket making.

Although it spreads too rapidly for many gardens, Santa Barbara sedge looks good as an informal, large-scale bank planting, and it forms a nice backdrop for large rocks or as a unifying understory for tough riparian herbs, shrubs, and trees. For a similar-sized plant but without the spreading habit, see *Carex mendocinensis*.

Carex bolanderi, Bolander's sedge, wood sedge
Prefers: mid- to high elevation, part shade, regular water
Blooms: early summer
Bolander's sedge is a fine-textured, slowly spreading sedge found over much of western North America in wetlands, moist meadows, and forest openings. Varying from ten inches to three feet tall, it is excellent as an accent in an irrigated landscape, as an understory for tall shrubs, or as an irrigated groundcover in light shade.

Carex brevicaulis, short-stem sedge
Prefers: low to high elevation, part shade, dry
Accepts: some summer irrigation
Blooms: early spring
Short-stem sedge is a small, dry-growing sedge that inhabits poor, rocky or sandy soils from coastal central California to British Columbia. It forms dense clumps of narrow leaves from two to eight inches tall, with flowerheads just slightly taller. The dense tussock is tough and makes an ideal understory for shrubs and trees, especially those that are sensitive to summer water. Plants can form a turflike meadow if planted close together.

Best in light shade, where, if grown dry, it retains more green into the summer, short-stem sedge accepts occasional summer irrigation. This is a good choice for sterile rocky soils such as serpentine, sandstone, or granitics.

Carex brevipes, see *C. rossii*

Carex capitata, capitate sedge
Prefers: high elevation, moisture, sun, good drainage
Accepts: low elevation and hot locations in part shade
Blooms: summer

This is a fine-textured sedge from marshy meadows to dry alpine slopes in the Cascades and Sierra Nevada to Alaska and South America. It forms loose clumps from four to fourteen inches tall with almost threadlike leaves and small flowerheads on wiry stems that rise just above the foliage. This is a lovely plant for moist sites with good drainage. Light shade is desirable in hot locations and at low elevations. Not often grown or available commercially, it is worth trying in gardens.

Carex concinnoides, northwestern sedge
Prefers: low to mid-elevation, part shade, regular water
Accepts: sun, heavy clay
Blooms: early spring
This is a dwarf sedge that spreads slowly from rhizomes, forming attractive mounds about four to twelve inches high, with deep, glossy green leaves that are sickle-shaped and deeply veined in the center. In early spring the flowers are produced above the foliage on wiry stems; yellow anthers top the spike of glistening white stigmas, which are translucent in the sunlight.

Native to the Klamath and North Coast Ranges to western Canada and Montana, northwestern sedge is best grown with some summer water and tolerates full sun or light shade and heavy clay soils. It is especially attractive in containers, where the floral details may be noticed, or as a small-scale groundcover in a rock garden.

Carex densa, dense sedge
Prefers: low to mid-elevation, sun, regular water
Blooms: spring
This is a medium-sized clumping sedge one foot to two feet tall, with flowers in dense clusters on wiry stems that rise above the foliage. Dense sedge grows at low to mid-elevations in seasonally wet areas in meadows and on slopes in the San Francisco Bay Area, the Cascade Ranges, the Sierra Nevada, the Central Valley, and coastal southern California to Washington and Nevada. Useful as an evergreen accent in meadows or as a low, tussocky groundcover at the margins of a woodland garden, it looks best with some summer water.

Carex echinata ssp. *phyllomanica*, star sedge
Prefers: low elevation, coastal, sun, regular water
Blooms: spring
A medium-sized to large clumping sedge about two to three feet across, star sedge has narrow, green or golden green leaves. It occurs in moist places at low eleva-

tions in scrub and forests of central and north coastal California to Alaska. It is attractive as a specimen plant in an irrigated garden or along the margins of a pond or lake. Wiry flower scapes are about four feet tall and have small greenish heads in an open panicle. **Carex echinata ssp. echinata** is similar and probably hardier, but this subspecies has rarely been tried in gardens.

Carex filifolia and **C. filifolia var. erostrata** are similar to *C. capitata* and have similar uses.

Carex fissuricola, cleft sedge
Prefers: mid- to high elevation, sun, regular water, good drainage
Accepts: light shade
Blooms: early spring
This is a choice medium-sized sedge growing twenty to thirty-two inches tall, with one-half- to two-inch-wide, strongly troughed, rich green leaves. The beautiful flowerheads appear in early spring and turn silky golden, staying presentable all summer. Cleft sedge is native to rocky soils in meadow and slope habitats and streamside at mid- to high elevations in the Sierra Nevada to the mountains of Idaho, Nevada, and Utah. It is easy to grow in sun or light shade and looks best with some summer irrigation. It is similar to both *Carex luzulifolia* and *C. luzulina*.

Carex globosa, round-fruit sedge
Prefers: low to mid-elevation, coastal habitats, part shade, regular water, good drainage
Accepts: some drought
Blooms: early spring
Native to well-drained soils in wooded areas near the coast in central and southern California, this sedge is six to twelve inches tall and spreads slowly to form bright green tussocks. Useful as a filler among shrubs or trees, especially in filtered shade, it is quite tolerant of drought but looks better with some summer water. It could be used as a groundcover in a lightly shaded meadow with only moderate foot traffic. Round-fruit sedge is similar to *Carex brevicaulis*, but spreads faster and has greener and slightly bolder foliage.

Carex gracilior, slender sedge
Prefers: low elevation, sun or light shade, regular water
Blooms: spring

This is an attractive clumping sedge, one foot to two feet tall. It is native to low-elevation, seasonally moist grasslands and slopes in open forests in the North, Central, and South Coast Ranges, the Sierra Nevada foothills, and the San Francisco Bay Area. Useful as an accent plant in sun or light shade, slender sedge looks best with regular irrigation during the dry season. It is similar in garden uses to *Carex densa*.

Carex luzulifolia is similar to *C. fissuricola* and has similar uses.

Carex luzulina is similar to *C. fissuricola* and has similar uses.

Carex mendocinensis, Mendocino sedge
Prefers: low to mid-elevation, shade, regular water, good drainage
Accepts: serpentine soils
Blooms: spring
This is a medium-sized clumping sedge native to moist areas, and often serpentine soils, in grasslands and forests in the North Coast Ranges, northern Sierra Nevada, and the Klamath Ranges into southern Oregon. Variable in size, depending on age and sterility of soil, it is small when juvenile or on poor soils, with broad, relatively short leaves. Older plants or plants grown on rich soils can be much larger, with narrow, half-inch-wide or wider leaves that are two or three feet long.

Flower scapes emerge in early spring from the semi-evergreen mound of foliage and are conspicuous when the male anthers are present, initially on eight-inch scapes and eventually elongating to two or three feet. Attractive as an accent plant, streamside, pondside, or in moist meadows, Mendocino sedge is best grown with summer water and good drainage.

A smaller (to one foot tall) golden-leaved form found by David McCrory in Sonoma County has been called "Cedars Gold."

Carex multicaulis, rush sedge, forest sedge
Prefers: low to mid-elevation, part shade, regular water, good drainage
Accepts: sterile soils, some dryness
Blooms: spring
Rush sedge is an attractive, delicate, medium-sized clumping sedge with evergreen, rushlike (round and wiry), arching leaves one foot to two feet tall. Native to woodlands and forests of the North and South Coast Ranges, the Sierra Nevada, and the Klamath Ranges, this elegant, fountainlike sedge deserves much wider use as

Carex nudata

an accent or groundcover in light shade, as a specimen in containers, or as textural contrast in rock gardens. It looks best with some supplemental water in summer, and it tolerates sterile soils.

Carex nudata, torrent sedge
Prefers: low to mid-elevation, part shade, wet places
Accepts: sun, periodic inundation
Blooms: late winter
Torrent sedge is a lovely deciduous sedge found in rocky or sandy streambeds below the high-water mark in the North and South Coast Ranges, the Sierra Nevada, the Sacramento Valley, and along the central California coast into Oregon. Its tough root system enables it to survive winter inundation and floodwaters well above the crown of the plant.

This is a tightly clumping sedge, and very old plants may build up a "stump"

Carex nudata. (CHARLES KENNARD)

several feet tall, with each year's growth emerging from the top. In late winter, as waters recede, torrent sedge sends out intricately beautiful inflorescences from the dormant crown of the plant, the black bracts highlighting the yellow pollen at the top and the glistening, translucent stigmas below. Soon new leaves emerge and form a lovely gray-green mound two to three feet across. With cooler autumn days, the foliage turns greenish yellow, then golden yellow, and finally tan-orange-gold as it floats languidly with the water's flow, seemingly in anticipation of the winter inundation that will strip the old leaves off, leaving dark stumps behind.

A marvelous sedge in its native habitats, torrent sedge also performs beautifully in irrigated gardens in sun or part shade. It is a natural for creekside plantings, especially as a dramatic clumping plant that anchors gravel, sand, or rocks along the shore and even into shallow waters. To promote root establishment in such sites it is best to plant seedlings as soon as rains stop and water recedes in spring. Although it seems to prefer moving water, torrent sedge thrives on the edges of pools or ponds or in gardens with regular irrigation. Soil need not be soggy, but should never be allowed to dry out.

Carex obnupta, coast sedge
Prefers: low elevation, coastal habitats, shade, wet places
Accepts: salinity
Blooms: spring
This is a tall, invasively spreading sedge occurring in moist to wet, often saline locations from coastal strand to coniferous forests along the central California coast into Washington, British Columbia, and beyond. In scale, appearance, growth habit, and uses, it is similar to *Carex barbarae*, though possibly less cold-hardy.

Carex pansa, sand-dune sedge
Prefers: low elevation, coastal habitats, sun, regular water, good drainage
Accepts: foot traffic, most soils, some dryness
Blooms: spring
Native to coastal strands in northern California to Washington, this low, ever-green, fine-leaved, spreading sedge may be the best sedge for a turf substitute. It tolerates foot traffic and many kinds of soil, fills in to form turf in a single season (planted as plugs about eight inches on center), requires only one or two mowings a season, is relatively soft, and accepts watering regimens ranging from regular to few. It can be left unmowed to form a hummocky meadow or used between step-ping stones, as a groundcover under open shrubs, or in pots.

Sand-dune sedge has been cultivated mostly in the Coast Ranges of Califor-nia, so gardeners in both colder and hotter climates should experiment before in-vesting in large quantities. A similar sedge, and probably hardier, is *Carex praegracilis*.

Carex phyllomanica, see *Carex echinata* ssp. *phyllomanica*

Carex praegracilis, clustered field sedge
Prefers: low to mid-elevation, sun or part shade, regular water
Accepts: alkalinity
Blooms: spring
This sedge is similar in appearance and in horticultural uses to *Carex pansa*, but its spread is slower. Native to coastal strands, moist meadows, scrub, and forests in many parts of California, it tolerates conditions away from the coast, but for best appearance, some summer irrigation is needed.

Carex raynoldsii, Raynold's sedge
Prefers: mid- to high elevation, sun or light shade, regular water, good drainage
Accepts: some dryness
Blooms: early summer
Raynold's sedge is low to medium-height, dense, and spreading, with grayish green leaves that are stiffly erect and eight to thirty inches tall. Native to high meadows, forests, and seeps in the Cascade and Klamath Ranges, the Sierra Nevada, and the Warner Mountains to British Columbia, it is useful as an informal, hummocky groundcover with regular watering and good drainage, in sun or light shade. The similar *Carex serratodens* probably is better for low-elevation gardens.

Carex rossii, Ross's sedge
Prefers: high elevation, sun, dry, good drainage
Accepts: lower elevations in part shade with some irrigation
Blooms: late spring
This small clumping sedge is found mostly in high-elevation dry forests, meadows, and open woodlands of the Cascade Ranges, the Sierra Nevada, the San Gabriel, San Bernardino, White, and Inyo Mountains, and into Colorado, Arizona, and Canada. From two to sixteen inches tall, it can be used as an attractive tussock in a rock garden planting, and it should make a useful groundcover in mid- to higher-elevation landscapes. At low elevations it benefits from light shade with some supplemental water.

Carex senta, swamp sedge
Prefers: mid-elevation, sun, wet places
Blooms: spring
This large, one-foot to three-foot-tall clumping sedge is native to streambanks, marshy areas, and meadows of the north and central Sierra Nevada, central and south coasts, southern Sacramento Valley, Channel Islands, and San Jacinto Mountains to Baja California and Arizona. It is similar to *Carex nudata* and has similar uses.

Carex serratodens, bifid sedge, two-tooth sedge
Prefers: low to mid-elevation, sun, regular water, good drainage
Accepts: serpentine soils
Blooms: spring
     This gray-blue-leaved sedge forms large, spreading colonies in moist meadows, seeps, and streambanks at low to mid-elevations in the North and South Coast Ranges, the Sierra Nevada, and the San Francisco Bay Area. Its spread is moderately paced, but it can form pure stands one to four feet tall and many yards wide in great age, giving a uniform color and pleasant undulating character to the landscape.
     Best in confined spaces or in pots, bifid sedge could be useful for erosion control on banks and in gullies, and it is tolerant of serpentine soils. It is best with regular water, but plants tend to go semidormant in winter regardless of watering regime.

Carex spissa, San Diego sedge, sawgrass sedge
Prefers: low to mid-elevation, sun or shade, regular water

Accepts: some dryness in shade

Blooms: spring

California's most beautiful large clumping sedge, this plant is native to perennial seepages, streams, canyons, and woodlands in central and southern California into Baja California. It has steely blue-gray leaves three to six feet long and an inch or more wide, with two ridges paralleling the

Carex spissa. (CAROL BORNSTEIN)

central trough, adding textural interest to the plant's strong color and form. It tolerates sun or considerable shade, although some shade is preferable if the plant is grown with only occasional water.

Beautiful as a specimen plant, San Diego sedge is effective with other bold perennials or planted in mass along watercourses or as an understory in riparian woodland. It is one of California's showiest sedges in flower, with long terminal brushes of soft yellow stamens in spring set above elongated female flowers that resemble baby corn and mature to tan seed clusters. The roots of this species were used in basketry by Native Americans of central California. It is similar to *Carex amplifolia*, but taller and without the tendency to spread.

Carex subfusca, brown sedge
Prefers: low to high elevation, sun or light shade, regular water
Accepts: drought
Blooms: spring

This low-growing sedge spreads underground and has proven useful as an informal turflike lawn substitute. It will tolerate considerable drought, retaining some green and growing slowly, but it looks best with regular water. Brown sedge is best cut back at the end of summer or even twice a year if routinely watered.

This is an attractive plant for hanging baskets or containers or as an informal groundcover in sun or light shade. It is native to meadows, chaparral, forests, and woodlands in many parts of northern, central, and southern California, including the desert mountains. Local or regional forms should be sought for best performance in the garden.

Carex teneriformis, see *C. subfusca*

Carex tumulicola, slender sedge, split-awn sedge
Prefers: mid-elevation, sun or shade, some summer water
Accepts: deep shade, some foot traffic, drought
Blooms: late winter, early spring
This attractive low, clumping sedge has narrow, bright green leaves throughout the year and scapes eighteen to thirty-two inches long that often grow outward,

Carex tumulicola. (SAXON HOLT)

making the plant look much lower. Native to meadows, forests, and open woodlands in the North and Central Coast Ranges, Sierra Nevada, San Francisco Bay Area, Channel Islands, and north to Washington, it is tolerant of full sun or deep shade and accepts drought, although it looks better with some summer water.

Slender sedge is especially useful as an informal groundcover under shrubs or trees that receive little water, and it is tough enough to survive some foot traffic. A form of this species collected by David Amme has been grown extensively and is called "Berkeley Sedge" by grass grower John Greenlee. It is commonly available in specialty nurseries.

Eleocharis acicularis* var. *occidentalis, needle spikerush
Prefers: low to mid-elevation, sun, wet places
Blooms: late winter, spring, early summer
Native to marshes, meadows, riverbanks, and vernal pools throughout much of California to the southeastern United States and northern Mexico, this perennial

spreading spikerush has wiry, eight-inch stems with small flowerheads at the tips. Widespread in vernally moist sites, it makes a loose, turflike, bright green groundcover.

Needle spikerush is attractive under taller, moisture-loving shrubs or open trees, especially where tough soil-binding is needed, such as along pond margins. Where spread needs to be controlled, it can be grown in containers.

Eleocharis bella is similar to *E. acicularis* var. *occidentalis*, but is an annual without spreading rhizomes.

Eleocharis montevidensis, Montevideo spikerush, sand spikerush
Prefers: low to mid-elevation, sun, wet places
Accepts: heavy clays
Blooms: spring
Montevideo spikerush has erect, bright green stems four to twenty inches tall, and spreads about three to six inches a year. It is native to moist meadows, seeps, and forests in parts of northern California, southern California, Texas, and South America. In gardens it is used much like *Eleocharis acicularis*, for areas of poor drainage such as heavy clay or along wet margins of lakes or pools. Its delicate, vertical growth tipped by tiny flower clusters makes it elegant displayed in containers. Plants remain bright green through the winter.

Eleocharis parishii, Parish's spikerush
Prefers: low to mid-elevation, sun, wet places
Accepts: foot traffic
Blooms: summer
Parish's spikerush is four to twelve inches tall and spreads underground. Native to moist, often sandy open areas in parts of northern and southern California to Oregon, Nevada, New Mexico, and northern Mexico, it is similar to *Eleocharis montevidensis* and has similar garden uses. It is useful for water margins and tolerates foot traffic and trampling by ducks and geese.

Eleocharis rostellata, beaked spikerush, walking sedge
Prefers: low to high-elevation, alkaline or saline soils, sun, wet places
Blooms: summer
Beaked spikerush is a tufted perennial with evergreen, one-foot to four-foot stems of vibrant green, the longer ones arching over and rooting at the tips. These arched

stems make a tight groundcover, difficult to walk through without tripping. It is useful for erosion control, especially in areas of alkaline or brackish water, and it can be grown in pots without drainage, as it prefers boggy soil, even standing water. It is attractive in hanging baskets. Roots are shallow. Plants spread by seed and rhizomes. Spread can be controlled in the open ground if the area above the upper limits of a water feature is kept dry in a band of ten feet or more.

Beaked spikerush is native to salt marshes, alkaline sinks, brackish seeps, thermal springs, and tidal flats from southern Canada south through Mexico to the West Indies, the Caribbean, and South America.

Eriophorum criniger, cotton grass
Prefers: mid- to high elevation, part shade, regular water, good drainage
Blooms: summer
Cotton grass is a beautiful sedge growing eight to forty inches tall in slowly spreading clumps of yellowish green foliage, topped by showy, whitish seedheads that resemble small fluffy pompons. Native to wet meadows and streambanks in northern and central California, the Cascade Ranges, and the Sierra Nevada to southern Oregon, it grows best in well-drained soils with regular water and prefers to stay on the cool side. An elegant accent plant for the moist rock garden or meadow or at the margins of a water feature, cotton grass would be especially useful in high-elevation gardens.

Isolepis cernua, club-rush, fiber-optics grass
Prefers: low elevation, coastal, full sun to light shade, regular water
Accepts: salinity
Blooms: late spring
Club-rush is a small clumping sedge, usually eight to twelve inches tall and twelve inches across, found in moist soils at lower elevations along the coast throughout California into British Columbia and South America. It can be either an annual or a short-lived perennial and is quick to form full-sized plants from seedlings or divisions. In cultivation it needs annual trimming back to keep growing vigorously, and growth will suffer if surrounding plants are competitive.

Club-rush is effective as a small-scale accent with water features, as a delicate groundcover, as an accent in a pot or hanging basket, or with other diminutive, moisture-loving perennials. It grows best with regular water, and tolerates full sun to light shade. Although plants of this species are commonly offered in nurseries today, these likely are not indigenous to California.

Schoenoplectus acutus **var. *occidentalis***, common tule or giant bulrush
Prefers: low to mid-elevation, sun, wet places
Blooms: summer

Common tule is a large, bold plant that spreads rapidly underground. It produces vertical, inch-thick, round, leafless stems to twelve or fifteen feet tall and can form massive colonies in the shallow fringes of wet areas. It is native to freshwater marshes, lakes, and streambanks throughout lower-elevation California, except the deserts, and much of temperate North America. Common tule and similar species were among the most important plants for Native Americans, who employed them in dozens of ways.

Although too large for all but the largest wet landscapes, common tule can be grown in small gardens in containers set just below water level, where its dramatic vertical form makes a powerful visual statement. In large landscapes, it can be used as a pond or river margin plant, where the tall stems and dense thickets attract wildlife and serve as perches for birds.

Common tule also can survive in ponds or pools that dry up in late summer, but will look better with regular water. In winter most stems turn brown, except in the mildest climates, where some green stems may remain. It is tidier if cut back to water level in winter or spring, although this will decrease its value for wildlife.

Schoenoplectus tabernaemontani, mountain bulrush
Prefers: low elevation, sun, wet places
Accepts: some dryness
Blooms: early summer

This four- to twelve-foot-tall tule, though generally smaller in most features than *Schoenoplectus acutus* var. *occidentalis*, is likewise best used as a containerized water plant. It can be grown in drier soils, which may control growth to some degree, but plants will perform best along open pond, lake, or river margins.

Mountain bulrush is native to freshwater marshes, lakes, and streambanks in coastal northern and southern California and the San Joaquin Valley to many parts of North America, South America, Eurasia, and the South Pacific. Plants from Riverside County are said to be grayish green rather than the more typical dull green. Several variegated cultivars are found in the nursery trade, but they are not likely to have originated in California.

Scirpus acutus **var. *occidentalis***, see *Schoenoplectus acutus* var. *occidentalis*

Scirpus cernuus, see *Isolepis cernua*

Scirpus criniger, see *Eriophorum criniger*

Scirpus microcarpus, mountain bog bulrush
Prefers: low to mid-elevation, sun, wet places
Blooms: spring, early summer
This is a large, bold, wet-growing, running sedge, two to five feet tall, that forms spreading clumps of glossy green, troughed leaves. The robust umbels of tan-white flowers are conspicuous and striking. Native to marshes, wet meadows, streambanks, and pond margins throughout California (except the deserts) to Alaska, the eastern United States, and Asia, this is an excellent plant for naturalizing around lakes, ponds, or riversides, though it is quite aggressive and can overwhelm more delicate plants. Fully deciduous in cold climates and partially evergreen in milder areas, it is attractive displayed in a water container.

Scirpus robustus, bull tule, big bulrush
Prefers: low elevation, coastal habitats, sun, wet places
Accepts: salinity
Blooms: early summer
Bull tule is a two- to five-foot-tall running sedge with distinctive triangular stems and troughed leaves. Common in low-elevation coastal saline and freshwater marshes in the San Francisco Bay Area and southern California to Mexico and South America, it makes handsome stands in large water features. In small gardens it can be displayed in water containers, where its interesting architectural character can be appreciated close at hand.

Scirpus tabernaemontani, see *Schoenoplectus tabernaemontani*

RUSHES

The rush family, Juncaceae, consists of eight or nine genera and over 300 species worldwide. California has only two genera, *Juncus*, rushes, with nearly sixty species, and *Luzula*, wood rushes, with six species. Much like the sedges, many rushes are associated with moist or wet places such as seeps or stream edges, pond margins, and marshes. *Luzula* species, unlike *Juncus*, do not tolerate soggy soil, but prefer moist, well-drained, woodsy conditions.

Most rushes in the genus *Juncus* are similar in appearance to many of the sedges in the genera *Scirpus* (tules) and *Eleocharis* (spikerushes). *Juncus* species have stiff, wiry, or cordlike green stems in crowded stands. In some species the leaves are reduced to inconspicuous scales or sheaths; others have narrow, irislike blades. *Luzula* species have broad grasslike or sedgelike leaves in small rosettes. Most *Juncus* species have tough rootstocks and can form either dense clumps or spread into large stands. *Luzula* species form small clumps or solitary plants with rather shallow roots.

Even when water is not visible, clumps of rushes are frequently indicative of seasonal moisture or water beneath the soil surface. It is this natural linkage between moisture and rushes that makes these plants so valuable to accent, or merely to suggest, the presence of water in the garden. Although many rushes occur naturally in or near water, most tolerate regular garden watering, and several look decent with little summer irrigation.

Rush flowers usually are borne in small clusters or sprays and, like those of the grasses and sedges, are composed of tiny parts that are fascinating to observe at close range and beautiful when the anthers and pistils are fertile. From a distance most rush flower clusters appear greenish, russet, or brown. The small, brown seed capsules may remain on the plant for many months and are an attractive accent, especially on the wiry-stemmed species.

Rushes were used by some Native American tribes for food, medicine, building material, mats, clothing, dye, ceremonial objects, and, most commonly, in coiled baskets. In basketry, rushes were frequently combined with sedges and grasses and other plant parts to create the finished product. Rush stems or roots were processed by drying, splitting, scraping, and dyeing before use or storage for later use.

Growing Rushes

Rushes are easily grown from seed or by division. Seeds can be sown from fall through spring, although seeds of most species will not germinate until warmer weather arrives. When plants have developed three to five leaves, they can be transplanted to small pots or flats for growing on to larger size. Divisions are best taken between fall and spring, especially if they are to be transplanted directly into the ground. Regardless of when they are divided, they must be kept moist until new growth is well under way, and watering should be continued regularly throughout the first season.

In the landscape, rushes are used much like grasses and sedges, except that most rushes need more supplemental watering. Many make excellent accent plants, and they can be massed for a grand effect. Like sedges and grasses, rushes may be clumping or they may spread from underground runners. The spreading forms are best used in large landscapes, where they are especially useful for erosion control or stabilization; in smaller gardens they can be grown in containers or confined by underground barriers.

Even the clumping forms can seed aggressively and become a nuisance. Self-sown seedlings prefer open soil, so heavy mulching can mitigate the problem. If unwanted seedlings appear, they should be removed when young, as even small plants can have tough root systems that make them difficult to eradicate.

Unlike most grasses, and closer to sedges in this regard, rushes do not respond well to mowing, except for yearly trimming during the dormant season. While some species tolerate more frequent cutting, most are slow to revive, and health and vigor decline if plants are cut back frequently. A good balance may be achieved with more aggressive species, but most rushes take considerable time to regenerate.

Rushes for the Garden

The following rushes are some of the best candidates for garden cultivation, and they are reasonably available in the nursery trade.

***Juncus acutus* ssp. *leopoldii*,** Leopold's rush, southwestern spiny rush
Prefers: low elevation, coastal habitats, alkaline conditions, sun, wet places
Blooms: summer
Leopold's rush is a large, nearly spherical plant, four to five feet tall and about seven feet across. It makes a beautiful accent and is dramatic when viewed from a distance. The stems are sharp-pointed at the tips, making it essential to place this plant out of the way of passersby. This rush is native to coastal salt marshes, alkaline seeps, and freshwater marshes in desert mountains in southern California, the southern Channel Islands, and the Sonoran Desert to Arizona, northern Baja California, South America, and southern Africa. It is tolerant of alkaline soils and looks best with regular watering. Because of its limited distribution in California, it is on the CNPS "watch" list (List 4). Seeds are available from specialty nurseries and seed suppliers.

Cutting back the plant in late winter to eliminate dead material may benefit its appearance, but where moisture is sufficient, renewal may be not be needed for many years. The large size and strong architectural form of this plant make it particularly useful in large landscapes or in association with other large plants, such as

California fan palm (*Washingtonia filifera*), and it would look natural in an arroyo (dry streambed) planting design. The attractive and conspicuous clusters of orange-brown seedheads project beyond the stems. It is best grown from seed, or plants can be divided after cutting back the spiny foliage.

Juncus bolanderi, Bolander's rush
Prefers: low to mid-elevation, sun, wet places
Blooms: spring
Bolander's rush is a spreading rush, twelve to thirty-two inches tall, with round, wiry stems and decorative brown seedheads in tight balls. Native to swampy or sandy ground over a wide area of the Northwest, it is useful as a soil binder at the edge of a pond or in containers as an accent, either in shallow water or irrigated regularly.

Juncus covillei, Coville's rush, common bog rush
Prefers: low to mid-elevation, coastal habitats, sun, regular water
Blooms: early summer
Coville's rush is a flat-leafed rush that spreads underground. Native to moist places, especially in forests, in the San Bernardino and San Gabriel Mountains, the Sierra Nevada and Cascade Ranges, and throughout much of the Northwest to British Columbia, it grows to about ten inches tall with showy seedheads in open, branched panicles. It forms dense, bright green drifts and can be used like unmown grasses in moist areas or grown with regular irrigation; it is winter-deciduous. It is effective in containers, where plants stay only a few inches tall.

Juncus effusus var. *brunneus*, soft rush, common bog rush
Prefers: low elevation, coastal habitats, sun, moist places
Blooms: late spring, summer
Soft rush is a large, elegant clumping rush from two to almost five feet tall with bright green stems. Native to low-elevation moist places, including salt marshes, throughout California (except the deserts) north to British Columbia, it grows well in gardens with regular water. This is an excellent, tough plant for water margins, as the bold clumps make a strong accent and the roots have exceptional soil-binding capacity.

   Although associated with water, soft rush is not an aquatic, and the crown of the plant should always be above water level. This is a wonderful plant for containers, where its presence suggests water. In the open ground, the clump spreads slowly outward, eventually forming a colony of clumps (the center dies out). The bright green color contrasts nicely with the grayish colors of *Juncus patens*, which is slightly smaller, or it can be paired with a clumping sedge such as *Carex spissa*. Plants from coastal northwestern California have especially fine foliage, but none has yet been selected and named.

Although little is known of their culture, **Juncus effusus var. gracilis** and **J. effusus var. exiguus** are worth trying in higher-elevation gardens. A number of cultivars may be found in the horticultural trade, but none is believed to be of California origin.

Juncus lesueurii, salt rush, dune rush

Prefers: low elevation, coastal habitats, sun, wet places, dunes
Accepts: salinity
Blooms: summer

Salt rush is rapidly spreading, with sinuous, cordlike stems one foot to three feet long and clusters of brownish flowers that are surprisingly showy when the iridescent pink stigmas and yellow anthers are produced. A wet grower from coastal dunes and salt or freshwater marshes in northern and central coastal California, it is useful as a stabilizer of soils at water margins, or it could be grown in a container with regular irrigation.

Juncus oxymeris, pointed rush

Prefers: low to high elevation, sun, some moisture
Blooms: summer

This is a moderately spreading rush, one foot to two feet tall in loose clumps, with handsome soft green, hollow, round stems and open umbels of medium-brown flowers. It holds its green color best with occasional irrigation, and it is useful as a soil binder near water. Native to montane meadows, shrubland, and forests, mostly at mid- to high elevations, in the Inner North Coast Ranges, Sierra Nevada, and Transverse Ranges to Washington, it is winter-deciduous except in the mildest areas.

Juncus patens, common rush, California gray rush

Prefers: low to mid-elevation, moisture, sun or light shade
Accepts: some dryness, foot traffic
Blooms: early summer, summer

Common rush grows sixteen to thirty inches tall in a slowly spreading clump. A wide-spreading plant with wiry, dark grayish-green stems, it is native to many plant communities in southern and central coastal California, the Channel Islands, and the Klamath and North Coast Ranges to Oregon.

Prone to self-sowing in moist or wet habitats, common rush perhaps is best grown in dryish landscapes, although it will go dormant or deteriorate without any summer irrigation. An evergreen rush that is attractive year-round if kept somewhat moist, it is extremely tough, and will tolerate light grazing and considerable trampling. It is excellent for use in drifts to provide a low-maintenance groundcover in sun or light shade and also in containers alone or in combination with other plants.

Juncus patens 'Occidental Blue'. (SAXON HOLT)

'**Carman's Gray**' is an attractive steely gray selection from Ed Carman's nursery in Los Gatos named by John Greenlee.

'**Elk Blue**' is a bluish gray selection by Randy Baldwin found near Elk in Mendocino County.

'**Mt. Madonna**' is a blue-gray selection by Jeff Rosendale found on Mount Madonna in Santa Cruz County.

'**Occidental Blue**' is a compact and well-behaved blue-gray selection by Bob Hornback found near Occidental in Sonoma County.

Juncus phaeocephalus, brown-headed rush
Prefers: low to mid-elevation, sun, regular water
Blooms: late spring, summer
Brown-headed rush is native to moist places in coastal strand and coastal scrub, the northern Sierra, San Bernardino Mountains, northern Channel Islands, and Peninsular Ranges. It is a spreading plant, four to twenty inches tall, with flat leaves and open clusters of brown balls of flowers and seedheads. It is good for an infor-

mal turflike margin to wet areas, where its root system will help prevent erosion. It is also effective in containers set at or just below water level in a pool or pond. It can be grown with regular irrigation.

Although both spread by runners, **Juncus phaeocephalus var. paniculatus** is more clumping than **J. phaeocephalus var. phaeocephalus**, in which the foliage is more evenly spread out along the runners.

Juncus tenuis, path rush, poverty rush
Prefers: low to mid-elevation, sun, regular water
Accepts: wind
Blooms: late spring, summer, fall
Path rush is an elegant, fine-textured, clumping plant eight to twenty inches tall that makes airy, evergreen clumps with graceful, arching seedheads. A choice plant for a bright green textural accent, it is particularly good for windy sites and attractive grown in a hanging basket or other container. Native to damp places in northern and central California and throughout much of the world, it needs regular watering. Some forms are said to be short-lived, but plants self-sow.

Luzula comosa, Pacific wood rush
Prefers: low to high elevation, some moisture, light shade, good drainage
Blooms: late winter, spring, early summer

Luzula comosa

Pacific wood rush is a subtle yet lovely clumping plant, four to sixteen inches tall, with six-inch-long by half-inch-wide glossy green leaves, covered with silky hairs when young and later often tinted bronze or reddish. The soft foliage clump is topped by arching and widely branched inflorescences that extend beyond the leaves. The stems and inflorescences are often reddish tan or mahogany colored, and the small clusters of flowers and bracts that compose the inflorescence are especially attractive when the yellow anthers are extended at flowering.

Native to meadows, open woodlands, and coniferous forests throughout the West, usually in light shade and especially on northerly slopes, Pacific wood rush provides a nice accent in gardens where it is not crowded out by more aggressive plants. It looks best with occasional summer water.

CATTAILS AND BUR-REEDS

The cattail family, or Typhaceae, is a cosmopolitan family of about twenty-five to thirty species worldwide in two genera: *Typha*, cattail, and *Sparganium*, bur-reed. Both cattails and bur-reeds are perennial marsh or aquatic herbs that grow from creeping rhizomes, often forming large colonies in shallow water or at the water's edge. Both can be striking in a water garden or in a large water container, and they are fascinating reminders of the great diversity of forms that nature provides.

Out of flower, cattails and some of the taller bur-reeds may look alike, and they may be found growing together in their native habitats. In flower, they are easily differentiated. Cattail flowers develop on the brownish, sausage-shaped, spikelike inflorescences for which the plants are widely known; the flowers of bur-reeds are white or yellowish and spherical, and the fruits are spiny, green or brown globes. Cattails and bur-reeds both form separate male and female flowers, with the male (staminate) flowers positioned higher up on the spike and female (pistillate) flowers below. Bur-reed flowers grow on the sides of flowering stems, which bend at each point of attachment, giving them a zigzag appearance. Cattail flowers grow near the tops of tall, straight stems.

The leaves of some bur-reeds are less than a foot to two and a half feet long and submersed or floating on the water; taller bur-reeds have leaves that are six to eight feet long, upright, and, like cattails, stand erect above the surface of the water. Cattails and bur-reeds are ubiquitous in freshwater marshes and other wet places around the world. Dense stands of these plants create a rich habitat for a wide array of other plants, insects, amphibians, fish, mammals, and birds. In the

wild, entire communities of wildlife exist unseen within the thick, seemingly impenetrable stands.

Growing Cattails and Bur-Reeds

Gardeners should be wary of introducing cattails and bur-reeds into the garden,
especially near wildlands where standing water is present for much of the year.
These plants in general are robust and vigorous, and some species can be extremely
invasive.

Even strong concrete collars can be circumvented by the aggressive root systems of cattails, either over the top or underneath, or the collars can burst under
the pressure of a burgeoning root mass. The rhizomes also can puncture thin pond
liners. If rhizomes begin to escape containment, immediate action should be taken,
as attempting to rid a pond of cattail rhizomes is a challenge indeed. Cattails also
spread, colonize, and invade by prolific production of wind- and water-dispersed
seeds. Where content and unrestrained, cattails may take over and create an almost complete monoculture. If cattails are grown in a small pond, the rhizomes
should be fully contained in five- or fifteen-gallon pots before setting in the water.
In spring or early summer, clumps can be divided or pieces of rooted rhizomes may
be removed and replanted in additional pots.

Cattails and bur-reeds are easily propagated by seed or by division of rhizomes.
Seeds can be collected and dry-stratified, then sown in spring in pots in moist soil
in a shady location; seeds also will germinate under water. Plants can tolerate brief
periods of drying out, but they grow best in at least a foot of water.

Cattails and Bur-Reeds for the Garden

The three California *Typha* species are for practical purposes identical in culture
and similar in appearance, and they may be used interchangeably in gardens large
enough to support them. The three species hybridize. Hybrids between *T.
angustifolia* and *T. latifolia* have been called *T. x glauca*.

Two subspecies of bur-reed, *Sparganium emersum* ssp. *emersum* and *S.
eurycarpum* ssp. *eurycarpum*, are especially good candidates for the cultivated water
garden. The two species are native to many other parts of the world, and both are
available in the nursery trade, so native plant gardeners should ensure that the
California subspecies are the ones that are offered.

The following cattails and bur-reeds are good candidates for large water features and container plantings in areas where spread can be contained. They are occasionally to often available in the nursery trade.

Sparganium emersum ssp. *emersum*, emersed bur-reed
Prefers: low to high elevation, wetlands, shallow water, sun or part shade
Blooms: summer
This is a robust, bright green, strap-leaved plant to three feet tall, with unbranched inflorescences bearing the spherical male flowers with bright white stamens and greenish white female flowers characteristic of the genus; the female flowers develop into spiny, burlike fruits composed of many nutlets. Plants are found mostly in northern California in coastal wetlands and high-mountain lakes to 8,500 feet. Leaves are held erect above the surface of the water. Its compact size makes this a fine plant for water-filled containers, alone or in combination with other aquatics.

Sparganium eurycarpum ssp. *eurycarpum*, giant bur-reed
Prefers: low to mid-elevation, wetlands, shallow water, sun or part shade
Blooms: summer
Native to ponds, lake margins, marshes, and streams in many parts of North America and beyond, this bur-reed may reach six feet in height. The inflorescences are branched. When not in bloom, plants resemble cattails, but the flowers and fruits are distinctive. This is a fine plant for pond margins, and it may be combined with cattails to good effect in the water garden.

Typha angustifolia, narrow-leaved cattail
Prefers: low to mid-elevation, full sun, wet places
Accepts: brief periods of dryness, brackish water with freshwater seepage
Blooms: late spring, early to mid-summer
Narrow-leaved cattail occurs in brackish or freshwater wetlands at low to mid-elevations in the Sierra Nevada (Lake Tahoe), the Central Valley, the central and south coasts, and the San Francisco Bay Area. It is also found in eastern North America and Eurasia and may be naturalized in, rather than native to, California. In fact, it grows almost anywhere that ground remains wet, saturated, or flooded during the growing season, including irrigation canals and roadside ditches. Broad-leaved cattail (*Typha latifolia*) and narrow-leaved cattail are often found in the same area, with the latter growing in slightly deeper water.

For a small pond or a refined effect, narrow-leaved cattail would be preferred because of its relatively compact height (to six feet) and slender leaves (one-quarter to one-half inch wide). The leaves are deep green. The stems are slightly shorter than the leaves, and the flower spikes have a short gap of bare stem (one to three inches) between the upper (male or staminate) portion and the lower (female or pistillate) portion. The male portion of the spike is light brown; the female portion is brownish green, turning dark brown as seeds mature. At maturity, the spike bursts under dry conditions, releasing the fruits.

Typha domingensis, southern cattail
Prefers: low to mid-elevation, full sun to part shade, wet places
Accepts: brackish water with freshwater seepage
Blooms: early summer
Southern cattail is found worldwide in warm temperate and tropical locations; it occurs in freshwater wetlands throughout California. The yellowish green leaves are usually slightly wider than those of *Typha angustifolia*. Plants grow eight to ten feet tall, with leaves slightly taller than the flower stems. Pistillate (female) and staminate (male) flowers are separated by a three-quarter-inch gap of bare stem. The female portion of the spike is wider than the male and bright yellow to straw-colored or orange-brown; the male spike is cinnamon brown.

Typha latifolia, broad-leaved cattail, common cattail
Prefers: low to mid-elevation, sun, wet places
Blooms: early summer
Broad-leaved cattail has grayish green strap leaves to an inch wide and ten feet long, with stems standing slightly taller. The leaves are thick and somewhat spongy. The male flowers are brown; the female flowers are pale green, drying to brown and later turning blackish or reddish brown in fruit. The male and female sections of the spike are continuous or only slightly separated.

This is the most common and widespread cattail in California, where it is found in freshwater wetlands, wet meadows, seeps, and many other places with standing water. It is also found in other parts of temperate North America, Central America, Eurasia, and Africa. Easily grown but very invasive, broad-leaved cattail grows much too large for the average water garden, and it must be contained.

ENDNOTES

[1] M. R. Duvall et al., "Phylogenetic Hypotheses for the Monocotyledons Constructed from rbcL Data," *Annals of the Missouri Botanic Garden* 80 (1993): 607–19; Wendy B. Zomlefer, *Guide to Flowering Plant Families* (Chapel Hill: University of North Carolina Press, 1994).

[2] James C. Hickman (ed.), *The Jepson Manual: Higher Plants of California* (Berkeley: University of California Press, 1993); L. Watson and M. J. Dallwitz, *The Grass Genera of the World* (Wallingford, Eng.: C.A.B. International, 1992), pp. 1–3. 45–54; Zomlefer, *Guide to Flowering Plant Families*.

[3] Hickman (ed.), *Jepson Manual;* Zomlefer, *Guide to Flowering Plant Families*.

Agave shawii. (STEPHEN INGRAM)

IV. SUCCULENT AND XEROPHYTIC PLANTS

Some of the most distinctive California native monocots—the yuccas, agaves, and nolinas—are encountered in the arid and semiarid southern regions of the state, as well as in other parts of the Southwest and in Mexico. In the wild, they catch one's attention when dry slopes, ridges, and canyons light up in spring with an astonishing profusion of tall inflorescences bearing massive displays of white, greenish white, cream, or bright yellow flowers. Out of bloom they may fade into the chaparral or stand out only in silhouette against the spare background of rocky cliffs or desert sands. Large and robust, with fleshy or fibrous, often sharp-pointed, swordlike leaves clustered into rosettes, they are outstanding specimens in the garden and, where they are content, can survive for many years.

AGAVES, YUCCAS, AND NOLINAS

Agaves and yuccas, traditionally placed in the Liliaceae or lily family, are now widely considered to be members of the agave family (Agavaceae),[1] which consists of about eight to twelve genera and 200 to 300 species of plants, widespread in dry subtropical areas of the southwestern United States to South America. Nolinas, sometimes included in the Agavaceae[2] and formerly placed in the lily family, today are often placed in their own family, the Nolinaceae,[3] or in the Dracaenaceae[4] along with the tropical and subtropical dracaenas, sansevierias, and cordylines. However, many botanists and taxonomists still include all three groups of plants in the Liliaceae, and this is where they may be found in some of the most commonly referenced texts.[5]

Regardless of their taxonomic classification, agaves, yuccas, and nolinas share a number of traits that make them a natural grouping for gardeners. They have strong, architectural form and tough, fibrous tissue in leaves, stems, and flower stalks. Their flowers are borne in terminal inflorescences on tall stalks. Their narrow leaves, held in a rosette attached to a stout stem, have a waxy cuticle that protects them from desiccation. They prefer dryness, and are effective grown in a collection of drought-tolerant plants. Some are also good container subjects, although many eventually become rather large.

Despite their similarities, agaves, yuccas, and nolinas are easily differentiated. Leaves of most agaves are stiff, with fierce, stout terminal spines and marginal teeth, and are broader and shorter than those of nolinas or most yuccas, the leaves of which have no marginal teeth. Yuccas and nolinas may be confused when not in flower, but yucca leaves are stiff and have a terminal spine and threadlike marginal fibers, while nolina leaves are lax to slightly recurved, most lack a terminal spine, and most have no marginal fibers, although the edges are rough and can cut one's skin. Spines can inflict painful wounds, and plants with spines should be sited away from areas where people will come in contact with them.

California agaves, yuccas, and nolinas are found in southern parts of the state in the extreme climates of the high and low deserts, as well as above ocean waves along the coast, in the company of conifers in the mountains, and in foothill chaparral. They need warmth in summer, sharp drainage, and little water. For some species, cold is not a problem, although wet cold may result in rot. Once established in the garden, they are unlikely ever to need irrigation. In wet-winter areas they can be protected from nature's generosity by planting on slopes, berms, or mounds and surrounding the crown of the plant with coarse sand or gravel. They have been grown successfully in northern California, planted on raised mounds to facilitate runoff and covered with plastic sheets during winter rains.

Many native agaves, yuccas, and nolinas are rare, threatened, endangered, or of limited distribution in the wild. They are easily grown from seed, and seeds and seed-grown plants of some species are available from specialty nurseries and seed exchanges.

AGAVES

Agaves are among the most striking plants of arid and semiarid landscapes, whether natural or designed. Agaves flower on tall, branched or unbranched stalks that grow from the center of a rosette of green to blue or blue-gray succulent or semisucculent leaves. The flower stalks, which are massive compared to the plant on which they develop, grow rapidly once they begin to emerge, sometimes elongating at a rate of two to three inches a day. There are about 250 species of *Agave*, about half of which are native to southwestern North America.

It may be that the Greek word *agavos*, which means noble or admirable, was applied to this group of plants not just for the awesome size of the flower stalk but for their simplicity and boldness of design. *Agave ferox* of Mexico is the largest and most impressive of all. The most frequently cultivated agave is another Mexican native, the century plant (*A. americana*), often encountered in one of its variegated-leaf forms.

Native Americans used agaves as food, drink, soap, clothing, rope, needles and thread, paper, glue, weapons, medicines, and dye. Agave leaves are still a source of fiber used for rope, rugs, and baskets, and juice from mature plants is consumed both fresh and fermented (as mescal or tequila). Steroid drugs have been synthesized from extracts of some plants in the agave family. As one writer put it, "The uses of agaves are as many as the arts of man have found it convenient to devise."[6]

Growing Agaves

The common name century plant, which is often applied to all agaves, is not accurately descriptive; considerably less time than a hundred years is required for any agave to flower, eight to fifteen years being sufficient for some species in some locations.

The flowers, however, are the culmination of a long and beautiful process, one that does, to be sure, take time. The flowering stalk, which can reach over twenty feet in some species, emerges from the rosette in late winter. When the stalk has reached its ultimate height, side branches bearing flower buds unfold. The opening buds become floral cups laden with nectar, which provides a sticky shower to anyone jostling the stalk.

The individual rosette dies after flowering, but offsets or "pups" continue to grow, allowing progeny from a single seed to live many decades or even centuries. Offsets can be pried away and replanted, or the dead rosette can be removed, and the pups left to continue growing in place. Some gardeners leave the stalk in place until it topples, as a testament to the enormous generative power that sent it skyward. If the stalk is removed, it can be used as a lever to pry the dead rosette loose from the colony. In the wild, insects and small creatures burrow into the dead flower stalk, creating a protected abode. The stalks of larger agave species often serve as woodpecker nests.

Agaves need excellent drainage, especially where winters are wet. They are best planted on mounds in fast-draining soil. Plants can be covered in winter to help prevent freezing and rot.

California agaves occasionally may be found in nurseries specializing in succulent and xerophytic plants. They have been difficult garden subjects in the Central Valley and should be considered mostly for southern California gardens, although *Agave shawii* thrives in the Berkeley hills and has been grown successfully in Napa County.

Agaves for the Garden

The following agaves are reasonably available in the nursery trade and adaptable to garden cultivation where conditions are favorable.

Agave deserti, desert agave
Prefers: low to mid-elevation, sun, dry, excellent drainage, protection from frost

Blooms: spring

Desert agave, as its common name suggests, is best adapted to desert conditions. It is native to rocky or gravelly soils on hot, open slopes and in washes in desert scrub in the mountains of southern California to Arizona and northern Baja California. Concentrations of this plant can be found in the low deserts of southern California and in arroyos below the western slopes of the San Bernardino Mountains.

Desert agave spreads primarily by underground rhizomes and forms dense colonies in the wild. As spent rosettes die and decompose, new ones replace them, gradually forming a ring-shaped colony ten to twenty feet in diameter. Stemless dead rosettes and flower stalks punctuate the colonies, having expended their energy in flowering. Although less productive than vegetative spread, pollination does occur; the primary pollinator is the desert bat, with which desert agave has a finely tuned symbiotic relationship.

The glaucous gray or blue-gray leaves of desert agave are rigid and coarsely fibered, with a thick cuticle, up to sixteen inches long by four inches wide at the base and tipped with long, pink-lavender spines and widely spaced marginal teeth that turn ashen white in age. The plant is more compact than some other agaves, with each rosette about eighteen inches tall and wide. It is very slow growing, taking about twenty years to flower. The branched flower stalks (panicles) are five to fifteen feet tall, and the flowers are bright yellow. The stalks of both the desert agave and the Utah agave (*Agave utahensis*) are slender and disproportionately tall in relation to the size of the rosette. Desert agave needs protection in winter if grown in northern California.

Agave shawii, Shaw's agave

Prefers: low to mid-elevation, coastal habitats, sun, dry, excellent drainage
Blooms: fall to late winter

Shaw's agave is found mostly on coastal bluffs and open slopes in Baja California, crossing into California at San Diego. Its natural habitats are coastal sage scrub and maritime desert scrub, and it is abundant along the northern Baja California coast. In California, however, it is rare, endangered, and almost extirpated. Endangered in California but more common elsewhere, Shaw's agave is on the CNPS List 2. Seeds and seed-grown plants occasionally can be obtained from seed suppliers and specialty nurseries or botanic garden plant sales.

Shaw's agave can hold its own in any plant collection. It eventually develops a short trunk from which the handsome, ovate, deep blue-green to gray-green leaves arise. Mature plants are about eighteen inches to two feet tall and five feet wide. Leaves can be up to eighteen inches long by six inches wide, tipped with a chestnut-colored spine and highlighted by burnished red, hooked teeth along the margin. Tightly folded leaves form a central

Agave shawii. (STEVE JUNAK)

bud. In unfolding, each leaf makes an imprint on the adjacent unfolded leaf, creating rich and intriguing patterns.

Plants may take up to fifteen years to bloom. Bright yellow flowers and purplish buds and bracts are congested in the branched flower stalk, which may reach fifteen feet or more. This is the most suitable agave for gardens near the coast, although plants may be stressed by persistent cool fog on the north coast, and may eventually rot from excessive winter rain. Provided with sharp drainage, however, they have survived decades in the San Francisco Bay Area.

Agave utahensis, Utah agave

Prefers: low to mid-elevation, sun, dry, alkaline soil, good drainage
Blooms: late spring, early summer

Utah agave is found in creosote bush scrub and Joshua tree woodland in the desert mountains of southern California to Utah and Arizona. Uncommon and of limited distribution, it is on the CNPS "watch" list (List 4). Plants and seeds occasionally are available from specialty nurseries and seed suppliers.

Utah agave grows to about eighteen inches tall and slightly more than a foot wide. The glaucous blue-gray or green leaves are narrow and lance-shaped but curved inward at the tips, with brownish to ivory marginal teeth and a long terminal spine. The slender flowering spike, up to five feet tall, lacks the branches of the other two native agaves, and the deep yellow flowers grow directly on the stalk. The stalk is usually about five feet tall, but can reach eight feet.

Utah agave is the agave encountered in the Grand Canyon. It prefers a hot, bright, dry location with perhaps some lime in the soil, and in hot areas it may appreciate an occasional summer shower. It tolerates cold, but not wet cold, and makes an excellent container plant.

Agave utahensis **var.** *eborispina*, ivory-spined agave, is especially striking, with its gray-green leaves with large marginal teeth and ivory-colored terminal spines. Rare and endangered in California and elsewhere, it is on the CNPS List 1B, but it has been sporadically available in the nursery trade for decades.

Agave utahensis **var.** *nevadensis*, the smallest and most cold-resistant variety, has blue-gray leaves and prominent marginal teeth. It is on the CNPS "watch" list (List 4), and is occasionally available from specialty nurseries.

Yuccas

Yuccas are still known in some parts of the West by their former common names: Spanish bayonet, Spanish dagger, and soapweed or soaproot (they were used for soap). Today, except for the Joshua tree, they are more likely to be known as yuccas, at least in California.

The name yucca, erroneously applied to American yucca plants by an early writer, comes from *yuca*, the Haitian name for the unrelated manihot. One species of yucca indigenous to the West Indies has flower stalks with pulpy flesh that is used for food and resembles the manihot in texture. The other forty species of yucca are confined to North America, especially the arid Southwest, and Mexico.

Most yuccas are stemless or nearly stemless rosettes of tough, swordlike, spine-tipped leaves with distended bases clasping the stem or rootstock. The leaf margins shred into filamentous fibers that are especially attractive when backlit by the sun. The leaves remain on the plant many years after it dies. Several yuccas are trees, including two California natives. All have strong silhouettes.

Growing Yuccas

Often seen in collections of succulent plants, yuccas also make fine specimens and are good company for large-leaved subtropical plants. They look at home in a mixed border of drought-tolerant native shrubs and perennials. Where drainage is adequate, some can tolerate garden watering. They thrive in dry heat and endure cold. Some are native to areas of fairly cold winters.

Yuccas flower more frequently than agaves, and, with the prominent exception of *Yucca whipplei*, flowering generally does not signal the death of the rosette. Nevertheless, flowering does drain the plant's reserves, and in nature plants flower only every few years. Flowering is more frequent in cultivation. The flower stalks have short side branches bearing large, creamy white, bell-shaped flowers and large bracts, the whole often strongly tinged dull red-violet. Livestock favor their sweet succulence. Where grazing is steady, there is little or no regeneration, and if continued over long periods, plants will be extirpated.

Yucca flowers are intensely fragrant at night, attracting the yucca moth (*Tegeticula* sp.), which is necessary for pollination of natural populations of many yuccas. After depositing eggs in the ovary of the yucca flower, the moth smears pollen brought from other plants onto the stigma. The fertilized flowers produce large, three-chambered capsules with six rows of flat, shiny, black seeds. Most

seeds are consumed by the moth larvae, but a few are left to start a new genera-
tion of plants. Near its natural range, the moth can be encouraged by providing
yuccas upon which it can feed.

Yuccas are easy to propagate. Offsets root easily in dry sand or a sandy mix.
Seeds germinate readily. Flowers can be artificially pollinated by smearing pollen
from one plant onto the stigmas of another.

Yuccas for the Garden

The yuccas described below are good candidates for garden cultivation where
their special needs can be met.

Yucca baccata, banana yucca
Prefers: mid-elevation, sun, dry, good drainage
Blooms: spring
Banana yucca has broad, concave, grayish or blue-green leaves to three feet long
by two inches wide at the middle, broadening to four inches at the reddish base
and shredding at the leaf margin into coarse, curly fibers. The rosette is stemless
for a number of years, but older plants having many offsets can produce as much
as three feet of often leaning trunk. Plant clusters tend to be broader than tall.

The flowering stem, two and a half feet tall, has about fifteen branches bearing gorgeous white to yellowish flowers, which are more than four inches long and
hang down. Fruits, up to eight inches long, are fleshy and resemble bananas.

Banana yucca is native to canyons and dry plains, washes, and slopes in Joshua
tree woodland and chaparral in the eastern desert mountains of southern California into Utah and Texas. The hardiest of California's native yuccas, it can take
considerable cold. It is propagated by seed or by offsets.

Yucca brevifolia, Joshua tree
Prefers: mid-elevation, sun, dry, good drainage
Blooms: spring
Joshua tree is so prominent in the high-desert landscape of southern California
that its name has been given to a national park and to an entire plant community,
Joshua tree woodland. It is found throughout much of the Mojave Desert; indeed,
it is the signature plant of that desert, as the saguaro is of the Sonoran Desert. It
sometimes intergrades with pinyon pine and juniper in higher-elevation parts of its
range. Some stands can be found at 6,500 feet elevation, covered with snow. It

Yucca brevifolia

Yucca brevifolia. (SAXON HOLT)

occurs east to southwestern Utah and western Arizona.

Joshua tree is by far the largest plant to be seen in its demanding environment. Occasional giants fifty feet tall have been reported, but half that height is more usual. Longevity of individual plants has not been determined, but plants are slow growing and will not reach full size in a gardener's lifetime. Growth form is variable.

Joshua tree is single-trunked and unbranched when young. Older trees usually are forked and dense, and very old trees tend to be single-stemmed, branching into an open crown, although some forms sucker profusely and make dense, low colonies, not treelike at all. Mechanical damage to the growing tip will cause it to fork; flowering itself may cause forking. The disposition, direction, and form of its heavy, looping branches distinguish the Joshua tree from anything else in the plant kingdom.

The specific epithet *brevifolia* describes the leaves, which are shorter than those of other yuccas. They are twelve inches or less in length and clustered toward the ends of the branches. Dead leaves remain appressed to the trunk for many years. The dense, heavy clusters of greenish white, bell-shaped flowers are up to a foot wide and branched. Capsular fruits are about four inches long.

Varieties of *Yucca brevifolia* exist that, though no longer given scientific recognition, are still horticulturally significant. What was formerly called *Y. brevifolia* var. *jaegeriana*, from the eastern Mojave Desert and eastward, grows to about eight feet tall and is smaller in all its parts than the typical species. The plant branches within three feet of the ground. Plants from western Antelope Valley, formerly called *Y. brevifolia* var. *herbertii*, send up clumps of three- to fifteen-foot leafy stems from an underground rootstock. Clumps may be up to thirty feet wide.

Joshua tree does not succeed in the Central Valley. It has survived many years, though not vigorously, in protected locations in the San Francisco Bay Area.

Yucca schidigera, Mohave yucca
Prefers: low to high elevation, sun, dry, good drainage
Blooms: spring

Mohave yucca (the usual spelling of the common name differs from the desert in which it grows) is long lived, slow growing, and highly variable. This is a large plant, sometimes treelike, that can be fifteen to over twenty feet tall, but usually is less than half that tall. It is usually branched with a crown of foliage up to twenty feet across. Stiff leaves can exceed three feet in length and are an inch to an inch and a half wide and yellow-green, with tough, curling marginal fibers.

The short-stemmed flower stalk may exceed three feet in length with fifteen to twenty-five branches. The creamy white flowers, often tinged dark violet, are quite beautiful; about two inches across, they are borne in eighteen-inch clusters. Capsules are about three inches long.

Mohave yucca grows in chaparral and creosote bush scrub on rocky slopes, mesas, and even coastal sea bluffs in San Diego County. It also can be found in the southern Mojave and northwestern Sonoran Deserts to Nevada, Arizona, and northern Baja California. In Baja it is known as *datil*. There are stands of Mohave yucca with blue-green leaves near Morongo Valley in San Bernardino County. It is quite easy to grow in the San Francisco Bay Area, where it is reliable and impressive.

Yucca whipplei, our lord's candle, chaparral yucca
Prefers: low to mid-elevation, sun, dry, good drainage
Blooms: spring to summer

Yucca whipplei grows in several plant communities, including coastal sage scrub and creosote bush scrub, but it is most commonly seen on slopes in steep, rocky canyons and in chaparral from Monterey County and the southern Sierra Ne-

Yucca whipplei. (STEPHEN INGRAM)

vada into northern Baja California. One reason for its popularity may be that it grows where millions of people live or go for recreation. Another undoubtedly is its striking beauty, especially when it is in flower. Plants in bloom make one of the most spectacular wildflower displays in California.

A bristling hemisphere of twelve- to thirty-inch-long, dull green to bluish, narrow, swordlike leaves tipped with brown spines sits on a fibrous rootstock. Leaves lack the marginal fibers of other yuccas. A five- to eight-foot shaft, densely covered with short branches of variable length and disposition in its upper half, eventually emerges from the rootstock. In warm weather the shaft can elongate more than a foot a day. Clusters of flowers are ivory-colored, sometimes tinged dark violet.

Where this plant is common, there will be many in bloom in spring, especially following a wet winter. As with agaves, rosettes are monocarpic, that is, they flower once and then die. Unlike agaves, many plants of *Yucca whipplei* do not produce offsets, so it is necessary to replace the plant when it dies.

In the wild, flower stalks remain on the plant for many years, memorial to a glorious display and a cozy abode for many creatures. Its garden use is suggested by its natural companions: coulter pine or pinyon pine, coffee ferns, dudleyas, salvias, bush monkey-flowers, California sagebrush, morning glory, scarlet bugler and other penstemons, scrub oak, ceanothus, and manzanitas. It is best to plant it where annual weeding is not required, to avoid contact with the dense masses of sharply armed leaves.

A number of plants formerly recognized as subspecies of *Yucca whipplei* may account for the variability of plants sold under this name. Branched plants from the desert edges have been called ssp. *cespitosa*; late-branching coastal plants have been called spp. *intermedia*; large, unbranched plants from the San Gabriel and San Bernardino Mountains have been called ssp. *parishii*; and rhizomed colonial plants from the South Coast Ranges have been called spp. *percursa*. Plants of *Yucca whipplei* still may be offered under these names.

NOLINAS

Nolinas are less well known than agaves and yuccas, perhaps because the latter are more common and more widely distributed in the wild. Nolinas resemble yuccas, but their leaves are longer, thinner, and more flexible. The dense crown of linear, gray-green leaves, up to 200 per rosette, sharp-tipped but without the spines of yuccas, sits fountainlike atop simple or sparingly branched trunks. The nearly naked ivory flowering shaft soars upward, its numerous branches subtended by long white bracts. The branches bear many congested small flowers, most often with male and female flowers on different plants. Male flowers are ivory white; female flowers are ivory white for the first day or two and then take on a greenish cast. The hundreds of sweet-smelling flowers are visited by thousands of bees.

Growing Nolinas

For form, texture, habit, size, and felicitous detail, nolinas are among the most undercelebrated monocots. They prefer full sun and require excellent drainage and tolerate cold, even wet cold conditions in milder areas. Nolinas are xerophytic plants, but generally are not considered succulents.

Under suitable conditions, nolinas can lay claim to pride of place in the garden, but because some grow quite large, they must be sited with care. Unlike agaves and some yuccas, they don't die after flowering, but continue to grow and bloom every few years. Plants shed older leaves, leaving a clean, stout trunk. In much of their range they are known as beargrass, possibly because of their resemblance to *Xerophyllum tenax* (beargrass) of the Pacific Northwest.

Nolinas for the Garden

Nolinas are not widely available in the nursery trade, and several are rare, threatened, or endangered, primarily because of development and horticultural collecting. Plants and seeds of the two listed below are occasionally available from nursery-propagated stock at botanic garden plant sales and from specialty growers.

Nolina bigelovii, Bigelow's nolina, Bigelow's beargrass
Prefers: low to mid-elevation, sun, dry, rocky soils, excellent drainage
Blooms: spring, summer
Bigelow's nolina bears creamy white to greenish white flowers that dry to an attractive straw color. The narrow, leathery, bluish green leaves are three to four feet long, forming a dense rosette that begins at ground level and in old plants tops a much-branched trunk three to eight feet tall. Older leaves have shredding, fibrous, brown margins. The tall, densely branched inflorescence is five to twelve feet tall.

Bigelow's nolina is found in creosote bush scrub in rocky or gravelly soils, on slopes and ridges and in canyons at elevations of 500 to 5,000 feet in the desert mountains of southeastern California to southern Nevada, western Arizona, and northwestern Baja California. It tolerates considerable cold and needs little or no water once established. Its showy flowers and striking architectural form suggest prominent placement in the landscape.

Nolina parryi, Parry's nolina, Parry's beargrass
Prefers: mid-elevation, sun, dry, excellent drainage
Blooms: spring, early summer
Parry's nolina is spectacular even when not in flower, and it is majestic in age. Trunks up to six feet tall may branch, giving the plant striking form and character. Four- to six-foot-long leaves are thick, pallid green or yellow-green, and one-half inch to one and a half inches wide, minutely serrate on the margins. The inflorescence can attain twelve feet in height, bare for the bottom half; the yellowish or creamy white flowers make a stunning display.

This nolina is native to dry slopes and ridges in chaparral and coastal sage scrub at elevations up to 3,000 feet and in pinyon-juniper woodland in the desert mountains of southern California, including the San Bernardino Mountains, the Peninsular Ranges, and the Sierra Nevada, at elevations up to 5,000 feet. It is cold-hardy, likes full sun and excellent drainage, and takes little or no water once established. Plants should be kept as dry as possible during winter. Planting on mounds and keeping fallen leaves and other organic material away from the base of the plant may help to prevent winter rot. Parry's nolina is quite easy from seed, but seedlings should not be set out in the open garden for two or three years.

Nolina parryi is sometimes offered as *N. bigelovii* var. *parryi*. *Nolina parryi* ssp. *wolfii* is a smaller form of the species native to the western Transverse Ranges and the north and western Peninsular Ranges. It does not seem to be available in the trade.

Nolina cismontana, recently segregated from *N. parryi*, is on List 1B of the California Native Plant Society (rare and endangered in California and elsewhere) and does not seem to be available in the nursery trade. It is native to the mountains of Orange and San Diego Counties.

Nolina parryi. (STEPHEN INGRAM)

ENDNOTES

[1] R. M. Dahlgren et al., *The Families of the Monocotyledons* (Berlin: Springer-Verlag, 1985), pp. 157–61; R. F. Thorne, "Classification and Geography of the Flowering Plants," *Botanical Review* 58 (1992): 225–348; H. S. Gentry, *Agaves of Continental North America* (Tucson: University of Arizona Press, 1982).

[2] A. Cronquist, *An Integrated System of Classification of Flowering Plants* (New York: Columbia University Press, 1981).

[3] Dahlgren, *Families of the Monocotyledons.*

[4] Thorne, "Classification and Geography of Flowering Plants."

[5] J. C. Hickman (ed.), *The Jepson Manual: Higher Plants of California* (Berkeley: University of California Press, 1993).

[6] Gentry, *Agaves of Continental North America.*

Washingtonia filifera. (WILLIAM FOLLETTE)

V. PALMS

Widespread throughout the tropics and subtropics of the world, palms are the universal symbol of tropical islands and desert oases, evoking powerful images of exotic places and earlier times. Palms are distinguished by their single, unbranched trunk and terminal tuft of luxuriant green leaves. Majestic in age, stately even when young, palms impart a festive and theatrical air to residential, commercial, and public landscapes large enough to accommodate them.

Palms are evergreen monocots in the Arecaceae, or palm family. Though treelike, they do not form true wood. Their shallow, fibrous roots are adventitious, arising from the base of the stem or trunk. Palms rely on overlapping leaf bases and vascular bundles at the periphery of the stem to remain erect, as do cycads and tree ferns.

Worldwide, there are about 200 genera and 3,000 species of palms,[1] more than two-thirds of which grow in tropical rainforests. The frost-tender coconut palm (*Cocos nucifera*), from the South Pacific and South America, is the signature tree of tropical isles; it grows outdoors in southern Florida. The commercial date palm (*Phoenix dactylifera*), from North Africa, is the palm grown commercially in southern California. The Canary Island date palm (*P. canariensis*) is widely planted as a boulevard tree and in large residential and commercial landscapes in both southern and northern parts of the state.

Growing Palms

Because of their noble form and intricate beauty, palms are among the most highly valued landscape plants in tropical and subtropical regions of the world, and people in temperate climates have long grown or tried to grow them. Groups of palms can be seen from long distances, and traditionally they have been used to demarcate a roadway or announce a destination. They are unequaled as stately sentries planted in rows along urban boulevards or the entry drives of grand estates. They define spaces and unify diverse architectural elements. Palms can be used as a symmetrical component of a formal landscape design, or as a vertical accent in informal plantings, either as single specimens or in small groves.

The best time for planting palms is late spring to early summer, when soil has warmed and root growth is most active. In cooler climates such as coastal northern California, newly planted palms may take two or more years to become established. The site should be well drained, and the planting hole should be at least twice as wide but no deeper than the root ball. Sand may be added to improve drainage, but organic material should be avoided, as it will break down and may cause the palm to sink into the ground. New roots will emerge both above and below the root ball. The root crown and trunk should not be covered with soil.

Newly planted palms should be watered daily to keep the root ball constantly damp but not soggy. As roots begin to draw moisture from a larger volume of soil, waterings should be spaced further apart, perhaps once a week during the first summer. Sandy soils may require more frequent irrigation and clay soils less frequent. Palms grow naturally in moist locations or where the water table is high, and most do not tolerate much dryness. Most established palms do well with slow irrigation to a depth of two feet every couple of weeks in summer and monthly to every six weeks in winter if the soil dries out and there is no rain.

Yellow or brown leaves may be left on the plant or removed in late summer. Green leaves should not be removed, especially above the horizontal plane, as this may result in disease, frost damage, or physical damage to the bud or trunk. Wounds to the trunk are permanent. Established plants may respond well to annual fertilization with a product formulated for palms.

For most palms, propagation from seed is not difficult as long as a few basic requirements are met. Among the most important are fresh seed, good sanitation, proper medium, proper hydration, and adequate heat. The fresher the seeds are, the better the results. The fruit pulp frequently contains a growth inhibitor, and removing it before planting will improve results. To do so, soak the seeds in water for forty-eight to seventy-two hours, changing the water daily; rub against a fine-meshed screen, then rinse with water until the seeds are completely free of pulp.

Seeds can be germinated in a commercial mix of peat moss or sterile sphagnum moss mixed with an equal amount of perlite or vermiculite. Sand, wood chips, screened rock, or volcanic cinder can substitute for vermiculite or perlite. Whatever medium is used, it should drain extremely well. Lay out seeds on the surface of the medium and dust them with a commercial insecticide. Bury seeds in the medium to half the seed diameter and cover with a finely screened cinder thick enough that it will not wash away during watering. This top dressing dries out quickly and discourages the moss that grows on peat.

Seeds should be watered thoroughly, then allowed to dry out before watering again. Too much water can greatly reduce germination. Once seeds begin to germinate, they will require more frequent watering. Seeds should be kept above 75 degrees F.

Palms for the Garden

California has only one native palm, California fan palm (*Washingtonia filifera*), which is native to the deserts of southern California and Arizona and is successfully grown at low to mid-elevations in northern as well as southern California.

Washingtonia filifera, California fan palm, desert fan palm
Prefers: low to mid-elevation, full sun to part shade, good drainage, moist places
Accepts: seasonal flooding, some dryness, heat, dry winds
Blooms: late spring, early summer
California fan palm is native to desert regions of southeastern California, Arizona, and northern Baja California, where it grows in seeps, springs, streamsides, or where groundwater is accessible. Large groves of this palm in their natural setting are magnificent and breathtaking. In the past, they were a welcome sight to travelers and explorers crossing the desert, as they indicated the presence of water and a shaded place to camp. Several desert towns in southern California have taken their names from groups of these palms—Palm Springs, Twentynine Palms, Thousand Palms, and Palm Desert.

Native Americans selected California fan palm oases as village sites, and they used its flowers and fruits as food, leaves as thatching, and leaf fibers for clothing and baskets. Because of their historical use by Native Americans, California palm oases are important archaeological sites. Many palm sites have been destroyed or are threatened by groundwater pumping, which has lowered the water table. Invasive exotics that compete for water, especially saltcedar (*Tamarix ramosissima*), have replaced palms in some areas.

California fan palm and its close relative, Mexican fan palm (*Washingtonia robusta*), are widely cultivated in California and in other regions with a similar climate. Mexican fan palm grows taller than California fan palm and has a narrower

Washingtonia filifera

trunk. The two species hybridize, and many variations in appearance occur, even among commercially grown plants. Both species have been cultivated in California since the 1870s and 1880s.

California fan palm is moderate to fast growing, with a straight, thick trunk that, when mature, may be thirty to sixty feet tall, and a loose, open crown of dark green, glistening foliage with a spread of fifteen to twenty feet. This palm is adapted to heat and low humidity, but requires good drainage and occasional deep watering.

California fan palm has pendulous, grayish green, palmate or fan-shaped leaf blades up to six feet across, with long, threadlike fibers along the margins. Leaves die at the end of the summer growing season, remaining attached to the trunk. If old leaves are not removed, they will form a dry, brown shag or "skirt" from the green-leaved crown to the ground. The branched inflorescences are longer than the leaves and at first erect, later hanging down among the leaves. Yellow and white flowers are followed by round black fruits, technically drupes, each containing a single glossy, reddish brown seed. California fan palm regenerates only by seed; vegetative propagation does not occur.

California fan palm thrives in a wide range of soils and climates, and it is quite resistant to diseases and pests. It responds well to fertilizer during the summer growing season, and in some soils may appreciate potassium and magnesium supplements. It tolerates drought, but benefits from regular watering. It tolerates frost and freezes with only minor damage to foliage. Palms in fast-draining soil seem to be more cold-hardy.

Propagation is easy from seed, and seedlings, or even mature specimens, can be transplanted. California fan palm is available commercially both as young plants and as larger specimens.

ENDNOTES

[1] James C. Hickman (ed.), *The Jepson Manual: Higher Plants of California* (Berkeley: University of California Press, 1993).

Kristin Jakob, Jenny Fleming, and Nora Harlow, May 2003. (PHYLLIS M. FABER)

ABOUT THE AUTHORS

Jenny Fleming was a founding member of the California Native Plant Society (CNPS) in 1965 and has been active in chapter plant sales since their inception. She has held state and chapter CNPS offices and is a Fellow of the Society. Jenny's large home garden is devoted almost exclusively to native plants.

Linda Haymaker is a landscape designer and horticulturist. She has worked as a senior staff member at U.C. Berkeley's Blake Garden, for the East Bay Municipal Utility District on water conservation planning and design, and privately as a landscape design consultant.

Kristin Jakob is a freelance botanic illustrator whose work has graced two journals and three books. Her numerous commercial illustrations have included a poster, "Wildflowers of the Sierra Nevada," for the California Native Plant Society. Her work has been exhibited in California, London, and Pennsylvania and has won awards from *Art Direction* magazine and from the Royal Horticultural Society, London. She also serves as a nursery and garden consultant.

Ron Lutsko, Jr., is a landscape architect who uses native plants extensively in his work. His designs include public and private landscapes, among them the native plant garden at Strybing Arboretum, the Santa Barbara Botanic Garden demonstration garden, and the Lindsay Natural History Museum demonstration garden in Walnut Creek. Trained in horticulture as well as landscape architecture, he has taught at the University of California in Berkeley and Davis. He was a member of the Horticultural Council that contributed to *The Jepson Manual: Higher Plants of California*.

Elizabeth McClintock was a research associate of the University of California Herbarium and of the Santa Barbara Botanic Garden and curator of botany at the California Academy of Sciences. A Fellow of the California Native Plant Society, she co-authored *A Flora of San Bruno Mountain*, authored *Trees of Golden Gate Park and San Francisco*, and has contributed to numerous botanic and horticultural publications.

Roger Raiche was for twenty-two years curator of the native plant collection at the University of California Botanic Garden in Berkeley. His extensive botanic fieldwork in California resulted in the discovery of three new taxa named in his honor, as well as the introduction of many superior cultivars of native plants. He currently is partner in the landscape design-build firm of Planet Horticulture, and he owns and manages The Cedars, a private nature preserve in Sonoma County devoted to the appreciation, preservation, and understanding of California's unique serpentine flora.

Wayne Roderick was formerly director of the Regional Parks Botanic Garden at Tilden and curator of the native plant collection at the University of California Botanic Garden at Berkeley. A Fellow of the California Native Plant Society and a member of the Horticultural Council that contributed to *The Jepson Manual: Higher Plants of California*, he received numerous international awards and honors. He brought public awareness and protection to many sensitive habitats, and he contributed seeds of native plants to horticulturists and growers throughout the state and the world. He passed away on August 10, 2003, just as this book was going to press. He will be sorely missed.

Suzanne Schettler is an ecological consultant and habitat restoration contractor with an extensive background in landscaping and nursery work. She has been active in the California Native Plant Society for over thirty years, including three as state president, and has served on the governing board of the Arboretum Associates of the University of California, Santa Cruz.

Jacob Sigg, Fellow and former president of the California Native Plant Society, before retirement was gardener supervisor at Strybing Arboretum and Botanic Gardens. Previous to that he was curator of Strybing's Arthur Menzies Garden of California Native Plants. His published articles span horticultural and conservation subjects, and he served on the Horticultural Council that contributed to *The Jepson Manual: Higher Plants of California*.

Nevin Smith has been a nurseryman much of his life, and his Wintergreen Nursery was for thirteen years a channel through which special selections of native plants entered the nursery trade. He writes frequently for *Pacific Horticulture* and *Fremontia*, has received several horticultural awards, and was a member of the Horticultural Council that contributed to *The Jepson Manual: Higher Plants of California*.

Caroline Spiller propagated native plants for the Strybing Arboretum Society for eighteen years and for six years was in charge of propagation for its annual plant sale. She travels widely to observe plants in their native habitats.

GLOSSARY

acid, acidity. See *pH*.

acute. Sharp, pointed, with a terminal angle between forty-five and ninety degrees.

adventitious. Plant parts produced in an unusual position or time of development, as roots at stem bases or buds along a stem rather than in a leaf axil.

alkaline, alkalinity. See *pH*.

alpine. Above timberline.

anther. The pollen-forming part of a stamen, generally an elongate or rounded structure at the tip of the slender, threadlike filament.

awn. A bristlelike appendage attached to a plant part. In grass, the bristlelike tip of the glume or lemma.

axil. The angle between the stem axis and the leaf petiole.

basal. At or near the base of a plant or plant part.

blade. The flat, expanded portion of a leaf or petal.

bract. A modified, generally much reduced leaf at the base of a flower or inflorescence.

bulb. The short, often rounded or disk-shaped underground stem covered by fleshy overlapping leaf bases or scales, as at the base of an onion. The bulb functions as a food storage organ.

bulbil. A small, secondary bulb that forms above ground, usually in a leaf axil.

bulblet. An immature bulb formed at the base of a mature bulb.

bunchgrass. Perennial grass forming more or less well-defined clumps, in contrast to those grasses that spread by rhizomes or stolons.

calyx. Collective term for all sepals of a flower, whether separate or united, below the corolla.

cambium. A layer of tissue, one cell thick, formed between the wood and the bark of vascular plants, that is capable of developing new cells. Growth of new wood takes place in the cambium.

campanulate. Bell-shaped.

capsule. Dry, generally many-seeded fruit.

caudex. Short, more or less vertical stem of a perennial, at or beneath ground level.

chaparral. Xerophytic vegetation, common in California, characterized by low-growing, mostly small-leaved evergreen shrubs or small trees that form dense, often impenetrable thickets.

chlorophyll. Green pigment of plants associated with photosynthesis.

cismontane. Situated on the seaward side of the mountain(s).

clone. A group of genetically identical individuals originating from a single parent plant by vegetative (asexual) reproduction.

conifer, coniferous. Cone-bearing tree or shrub.

corm. A shortened, compressed, generally underground stem covered by dry, scalelike leaves.

cormel, cormlet. New corms that form around the old corm.

corolla. Collective term for petals.

cotyledon. First leaf or leaves of the embryo of a seed plant.

culm. The stem of a grass, sedge, or rush.

cultivar. Contraction of *cultivated variety*. Used in a broad sense for plants selected for particular morphological, physiological, chemical, and other characters. Applied only to plants in cultivation, not wild-growing ones. When reproduced, generally vegetatively, cultivars retain their distinguishing characters. Compare *selection*.

damping off. Sudden plant death in the seedling stage, usually due to soil-borne fungi, but may be from seed defects, temperature extremes, toxics, or excessive or deficient soil moisture.

deciduous. Plants that shed leaves or other parts at the end of a growing period, as leaves that drop in autumn or flower parts that fall after the flowering period. Compare *evergreen*.

dicotyledon, dicot. Member of the main subgroup of flowering plants, generally with two cotyledons. Compare *monocotyledon, monocot*.

division. The result of a vegetative propagation technique in which a portion of a plant is dug up and divided into smaller parts (divisions) to start new plants.

dormant, dormancy. A period of inactivity or rest during part of the year when active growth temporarily stops and aboveground plant parts may wither and die back.

drupe. A stone fruit having a hard, nutlike inner part surrounded by a fleshy or fibrous outer layer; a fruit with a single seed.

embryo. The developing immature plant contained in the seed of flowering plants prior to germination.

endemic. Native and restricted to a well-defined geographic area.

established. Settled, actively growing, and more or less permanently fixed at a site.

evergreen. Having persistent leaves for two or more growing seasons. Evergreen leaves generally drop over a period of time, in contrast to deciduous leaves, which are shed annually.

fall. One of three drooping outer segments of an iris flower, equivalent to the sepals; the outer whorl of tepals of an iris flower.

filament. Threadlike part of a stamen supporting the anther.

floret. Small, individual flower that, together with others, may form a dense inflorescence, as in those of the grass family.

genus, genera. Group of related species.

glabrous. Lacking hairs.

glaucous. Covered with a waxy bloom.

glume. Sterile bract at the base of a grass spikelet.

head. Dense, short inflorescence of generally sessile flowers.

herb, herbaceous. Plant with no persistent woody stem above ground.

hybrid. Offspring of a cross between two species or subspecies.

hypocotyl. Part of a plant embryo or seedling below the cotyledon.

imbricate. Having margins overlapping in a regular order.

inflorescence. Arrangement of flowers along a flowering stem.

keel. A central ridge along the back of any part of a plant, such as a leaf. In an iris, the two front, often lower, petals of the flower.

lanceolate. Lance-shaped; much longer than wide.

lemma. In grasses, the lower and generally larger of the pair of bracts at the base of a floret. Compare *palea*.

Mediterranean climate. One of the climatic regions of the world where rain falls principally in the cool season and the warm season is dry. There are five such regions, including the Mediterranean, western California, central Chile, southwestern Australia, and the Cape region of South Africa.

mesic. Characterized by, relating to, or requiring a moderate amount of water.

mesophytic. Structurally adapted for life and growth with a moderate amount of water.

monocotyledon, monocot. The smaller subgroup of flowering plants, generally with one cotyledon. Compare *dicotyledon, dicot*.

montane. Of the mountain(s); between foothills and subalpine.

neutral. Soil with pH of 7.0 on a scale of 0 to 14, indicating a moderate level of hydrogen ion activity. Most plants grow best in neutral to slightly acid soil.

node. Joint of a stem where a leaf or other structure is borne; the space between two nodes is called an internode.

offset. A young plant produced by the parent plant, usually at its base. In bulbs, a small bulb at the base of the parent bulb.

ovate. Egg-shaped.

palea. In grasses, the upper, and generally smaller, of a pair of bracts at the base of a floret. Compare *lemma*.

panicle. Branched inflorescence in which opening of the flowers begins with the lower portion and progresses upward.

pedicel. Stalk of a single flower in a flower cluster.

pendant, pendulous. More or less hanging.

perennial. Plant living three or more years.

perianth. Calyx and corolla collectively.

petal. One of the individual parts of the corolla.

petiole. Stalk of a leaf.

pH. Symbol denoting the relative concentration of hydrogen ions in a solution or soil, indicating acidity or alkalinity. Most plants grow best in neutral (pH = 7.0) or slightly acidic (pH < 7.0) soil.

pistil. Female reproductive organ in a flower, consisting of an ovary, style, and stigma.

raceme. Unbranched inflorescence with flowers borne on pedicels along a main axis and generally opening from below upward.

radicle. Rudimentary stem of a plant that supports the cotyledons in the seed and from which the root develops.

recurved. Curved downward or backward.

reflexed. Recurved or bent downward.

reticulate. Resembling a net.

rhizome. Subterranean rootlike stem, generally producing roots from its lower surface and shoots from its upper surface.

rootstock. Rootlike stem; elongated underground structure by which plant spreads.

rosette. Pattern of growth in which leaves radiate from a crown, usually close to the ground.

runner. Slender stolon; stem lying on the ground.

saline. Salty.

scale. Small, often dry, appressed leaves or bracts.

scape. Leafless flower-bearing stem arising from ground level.

scree. Relatively unstable, sloping accumulation of small rock fragments, often at a cliff base.

selection. A plant chosen for its superior characteristics. The plant may have arisen as a result of a breeding program, or it may be a chance seedling that appeared in cultivation or was brought into cultivation from the wild.

sepal. One of the individual parts of the calyx, whether fused or not, generally green and normally enclosing the inner parts of the flower.

series. A grouping that may include several species; a classification that results when a genus is divided into subgenera, a subgenus into sections, a section into subsections, and a subsection into series.

serpentine. Green, yellow, brown, or reddish soil or rock with high levels of magnesium and iron and low levels of calcium. Serpentine soil is toxic to many plants.

sessile. Lacking a stalk.

sheath. Tubular basal section of the leaf that encloses the stem, common in grasses, rushes, and sedges.

sodic. Containing sodium.

spadix. Floral spike with a fleshy or succulent axis, usually enclosed in a spathe.

spathe. Sheathing bract or pair of bracts enclosing an inflorescence, especially a spadix on the same axis.

species. Group of like individuals that interbreed and are reproductively isolated from other such groups. Closely related species make up a genus.

spike. Unbranched inflorescence to which flowers are directly attached, nearly always opening from bottom to top.

spikelet. Portion of the inflorescence of grasses and sedges; a unit of one or more flowers on a ministem; in grasses, everything from a pair of glumes upward.

stalk. Main or supporting axis of a stem of any organ, as of a leaf or flower.

stamen. Male pollen-producing organ of a flower.

standard. One of three inner, more or less erect segments of an iris flower, equivalent to petals.

stigma. Part of a pistil, usually at the tip, that receives pollen and on which pollen germinates.

stolon. Horizontal stem at or above the soil surface that gives rise to new plants at the tip or at nodes. A trailing shoot above ground, rooting at the nodes.

stoloniferous. Having stolons.

style. Slender, more or less elongated part of the ovary, connecting it with the stigma.

subalpine. Just below timberline, between montane and alpine.

subspecies. Subdivision of a species with its own geographic distribution, differing morphologically and genetically in minor characters from the species and from other subspecies; used by some botanists in the same sense as variety.

succulent. Having fleshy tissues that conserve moisture.

talus. Relatively stable, sloping accumulation of large rock fragments, often at a cliff base. Compare *scree*.

taxon. A group of genetically similar organisms that are classified together as a species, genus, family, etc.

tepal. Petal or sepal, segment of perianth; any of the subdivisions of a perianth not clearly differentiated into calyx and corolla.

throat. Expanded, fused portion above the tube of a flower with fused sepals or petals.

tuber. Short, thick, fleshy underground stem that serves as food storage organ.

tunic. Outer covering of some bulbs or corms.

tunicate. Having concentric layers or coats, as the bulb of an onion.

umbel. Inflorescence composed of several branches that radiate from a single point and are terminated by individual flowers or by secondary umbels.

variety. Subdivision of a species occurring naturally or produced by selective breeding.

whorl. A group of leaves or other structures arranged in a circle at a single node.

xeric. Characterized by, relating to, or requiring only a small amount of moisture.

xerophytic. Structurally adapted for life and growth with a limited water supply, especially by means of mechanisms that limit transpiration or provide for storage of water.

SITE PREFERENCES OF CALIFORNIA MONOCOTS

This table should be considered in the context of the description of each plant in the text, which contains important details that help in interpreting the information summarized here. Unless otherwise noted, *shade* means light shade or part-day shade, and *sun* means part sun or part-day sun. *Moist* means some moisture preferred in summer. *Dry* means little or no water in summer. *XX* indicates greater suitability for the conditions noted.

Plant	Sun Dry	Sun Moist	Shade Dry	Shade Moist	Comments
Achnatherum parishii	X				drainage
Achnatherum speciosum	X				drainage
Achnatherum thurberianum	X				drainage; higher elevation
Agave deserti	X				excellent drainage
Agave shawii	X				excellent drainage; coastal
Agave utahensis	X				drainage
Agave utahensis var. *eborispina*	X				drainage
Agave utahensis var. *nevadensis*	X				drainage; cold resistant
Agrostis hallii	X	XX	X	X	some moisture
Agrostis pallens	X		X		drainage
Agrostis scabra		X			regular water; higher elevations
Allium acuminatum	X				drainage
Allium amplectens	X				clay soils, serpentine
Allium cratericola	X				excellent drainage
Allium crispum	X				clay soils, serpentine

Plant	Sun Dry	Sun Moist	Shade Dry	Shade Moist	Comments
Allium dichlamydeum	X		X		coastal; drainage
Allium falcifolium	X				excellent drainage
Allium haematochiton	X		X		drainage
Allium hyalinum	XX	X			drainage; some summer water okay
Allium peninsulare			X		excellent drainage
Allium platycaule	X				excellent drainage; higher elevations
Allium praecox			X		drainage
Allium serra	X				excellent drainage
Allium siskiyouense	X		X		excellent drainage; higher elevations
Allium unifolium	XX	X	X	X	coastal
Allium validum		X			wet places; higher elevations
Aristida purpurea var. *purpurea*	X				drainage
Aristida purpurea var. *wrightii*	X				excellent drainage; heat
Bloomeria crocea	X				
Bothriochloa barbinodis	XX	X			drainage; heavy clay soil okay
Bouteloua curtipendula	X				drainage
Bouteloua gracilis	X				drainage; heat
Brodiaea californica	X				clay soils, serpentine
Brodiaea coronaria	X				vernally moist
Brodiaea elegans	XX	X			clay soils; some summer moisture okay

Plant	Sun Dry	Sun Moist	Shade Dry	Shade Moist	Comments
Brodiaea jolonensis	XX	X			clay soils
Brodiaea minor	X				vernally moist
Brodiaea purdyi	X				heat
Brodiaea stellaris	X				coastal
Brodiaea terrestris	X				coastal
Calamagrostis foliosa		X			drainage; regular water; coastal
Calamagrostis koelerioides		X			excellent drainage; higher elevations
Calamagrostis nutkaensis		X			drainage; regular water; coastal
Calamagrostis ophitidis		X	X	XX	excellent drainage; serpentine; higher elevations
Calamagrostis purpurascens		X			excellent drainage; high elevations; regular water
Calamagrostis rubescens		XX		X	drainage; some summer water
Calochortus albus			X		drainage
Calochortus amabilis			X		drainage
Calochortus amoenus			X		drainage
Calochortus catalinae	X				drainage; coastal
Calochortus clavatus	X				drainage
Calochortus coeruleus	X		X		drainage
Calochortus luteus	X				drainage; clay soils, serpentine

Plant	Sun Dry	Sun Moist	Shade Dry	Shade Moist	Comments
Calochortus macrocarpus	X				drainage; protection from winter rains
Calochortus monophyllus			X		drainage
Calochortus nudus				X	moist; cool
Calochortus splendens	X				drainage
Calochortus superbus	X				drainage
Calochortus tolmiei			X		sun near coast
Calochortus umbellatus			X		sun near coast; serpentine
Calochortus uniflorus	X	X			vernally moist; coastal
Calochortus venustus	X		X		drainage; higher elevations
Calochortus vestae	X				drainage
Calochortus weedii	X				drainage
Camassia quamash ssp. breviflora		X			higher elevations
Camassia quamash ssp. quamash		X			low and high elevations
Carex amplifolia		X		X	regular water
Carex barbarae		X		X	invasive
Carex bolanderi				X	regular water; higher elevations
Carex brevicaulis			X		
Carex capitata		XX		X	drainage; higher elevations
Carex concinnoides		X		XX	regular water

Plant	Sun Dry	Sun Moist	Shade Dry	Shade Moist	Comments
Carex densa		X			regular water
Carex echinata ssp. phyllomanica		XX		X	coastal; regular water
Carex filifolia		XX		X	drainage; higher elevations
Carex filifolia var. erostrata		XX		X	drainage; higher elevations
Carex fissuricola		XX		X	drainage; higher elevations
Carex globosa				X	drainage; coastal
Carex gracilior		X		X	regular water
Carex luzulifolia		XX		X	regular water
Carex luzulina		XX		X	regular water
Carex mendocinensis				X	drainage; regular water
Carex multicaulis				X	drainage; regular water
Carex nudata		X		XX	wet places
Carex obnupta				X	coastal; wet places; invasive
Carex pansa		X		X	coastal; drainage; regular water
Carex praegracilis		X		X	regular water
Carex raynoldsii		X		X	higher elevations; drainage; regular water
Carex rossii	X				high elevations; drainage
Carex senta		X			wet places

Plant	Sun Dry	Sun Moist	Shade Dry	Shade Moist	Comments
Carex serratodens		X			drainage; regular water; invasive
Carex spissa		X		X	regular water; some dryness in shade
Carex subfusca		X		X	regular water; some dryness okay
Carex tumulicola		X	X	X	some drought in shade okay
Chlorogalum parviflorum	XX		X		drainage; heavy clay if dry in summer
Chlorogalum pomeridianum	XX		X		drainage; coastal
Clintonia andrewsiana				X	coastal; moist; summer dryness okay
Clintonia uniflora				X	northern; summer dryness okay
Cypripedium californicum				X	excellent drainage; regular water
Cypripedium montanum		X		X	excellent drainage; high elevations
Danthonia californica	X	XX			coastal; cool; some dryness near coast okay
Danthonia californica var. *americana*	X	X			coastal; cool; higher elevations
Danthonia intermedia	X	X			drainage; higher elevations
Deschampsia cespitosa ssp. *cespitosa*		X			coastal; cool; regular water
Deschampsia cespitosa ssp. *holciformis*		X			coastal; regular water; some dryness near coast okay

Plant	Sun Dry	Sun Moist	Shade Dry	Shade Moist	Comments
Deschampsia danthonioides		X			vernally moist; annual
Deschampsia elongata		X		X	higher elevations; drainage
Dichelostemma capitatum	X				drainage
Dichelostemma congestum	XX		X		drainage
Dichelostemma ida-maia	X		XX		drainage; some summer water okay
Dichelostemma multiflorum	X				drainage
Dichelostemma volubile	X		X		drainage; accepts some summer water
Disporum hookeri				X	drainage; some dryness okay
Disporum smithii				X	
Distichlis spicata		X			salt, alkaline marsh
Eleocharis acicularis var. *occidentalis*		X		X	wet places; invasive
Eleocharis bella		X		X	wet places; annual
Eleocharis montevidensis		X			wet places; poor drainage okay
Eleocharis parishii		X			wet places; foot traffic okay
Eleocharis rostellata		X			wet places; alkaline or saline soils; invasive
Elymus californicus				X	coastal; cool; sun near coast with water
Elymus elymoides ssp. *elymoides*	X				drainage; higher elevations; heat, drought okay

Plant	Sun Dry	Sun Moist	Shade Dry	Shade Moist	Comments
Elymus glaucus	X	X			drainage; invasive
Elymus multisetus	X				drainage; invasive
Epipactis gigantea		X		XX	wet places; drainage
Eriophorum criniger				X	drainage; cool; wet places; higher elevations
Erythronium californicum			XX	X	drainage; summer dryness okay
Erythronium citrinum			X		drainage
Erythronium helenae			X		drainage; serpentine
Erythronium hendersonii			X		drainage
Erythronium multiscapoideum	X		X		drainage
Erythronium revolutum				X	drainage; some summer water
Erythronium tuolumnense				X	drainage; some summer water
Festuca brachyphylla		X		X	excellent drainage; regular water; higher elevations
Festuca californica		X		XX	full sun with regular water okay
Festuca idahoensis		X		X	drainage
Festuca occidentalis				X	drainage
Festuca rubra		X		XX	regular water
Fritillaria affinis			X		drainage; sun near coast
Fritillaria biflora	X				drainage; coastal; serpentine
Fritillaria eastwoodiae			X		drainage

Plant	Sun Dry	Sun Moist	Shade Dry	Shade Moist	Comments
Fritillaria purdyi	X				drainage; serpentine
Goodyera oblongifolia				X	drainage; acid soil
Hastingsia alba		X			higher elevations; wet places; some dryness in shade okay
Hesperostipa comata	X				drainage; higher elevations
Hierochloe occidentalis				X	drainage; rich soils
Hordeum brachyantherum		X			vernally wet; heavy clay okay
Iris bracteata		X		X	drainage; occasional summer water
Iris chrysophylla		X		X	drainage; occasional summer water
Iris douglasiana	X		X		drainage; coastal; infrequent summer water okay
Iris fernaldii	X		X		drainage
Iris hartwegii	X		X		drainage; higher elevations
Iris innominata			X		excellent drainage; inland; some summer water near coast okay
Iris longipetala		X			drainage; coastal; occasional summer water
Iris macrosiphon	X				drainage; drought okay
Iris missouriensis		X			drainage; wet places; higher elevations

Plant	Sun Dry	Sun Moist	Shade Dry	Shade Moist	Comments
Iris munzii	X	X	X		drainage; inland; drought okay
Iris purdyi	X		X	X	drainage; drought okay
Iris tenax ssp. *klamathensis*	X		XX		excellent drainage; northern; infrequent summer water
Iris tenuissima	X		XX		drainage; drought okay
Isolepsis cernua		X		X	coastal; wet places
Juncus acutus spp. *leopoldii*		X			wet places; coastal
Juncus bolanderi		X			wet places
Juncus covillei		X			coastal; regular water
Juncus effusus var. *brunneus*		X			regular water; coastal; saline soils okay
Juncus lesueurii		X			wet places; coastal; saline soils okay
Juncus oxymeris		X			regular water; higher elevations
Juncus patens		X		X	regular water; some dryness okay
Juncus phaeocephalus		X			regular water
Juncus phaeocephalus var. *paniculatus*		X			regular water
Juncus phaeocephalus var. *phaeocephalus*		X			regular water
Juncus tenuis		X			regular water
Koeleria macrantha	X				drainage
Leymus cinereus	X				drainage; heat

Plant	Sun Dry	Sun Moist	Shade Dry	Shade Moist	Comments
Leymus condensatus	X	X			drainage; invasive
Leymus mollis		X			drainage; invasive
Leymus triticoides		X			invasive
Lilium humboldtii			XX		excellent drainage; dry summer dormancy
Lilium humboldtii ssp. *humboldtii*			X		higher elevations
Lilium humboldtii ssp. *ocellatum*			X		
Lilium kelleyanum				XX	drainage; wet places; high elevations
Lilium pardalinum		X		XX	drainage; full sun with water near coast
Lilium pardalinum ssp. *pitkinense*				X	
Lilium pardalinum ssp. *wigginsii*				X	excellent drainage
Lilium parvum				X	drainage; high elevations
Lilium parvum var. *luteum*				X	drainage; high elevations
Luzula comosa				X	drainage
Lysichiton americanus		X		X	coastal
Maianthemum dilatatum				X	coastal
Melica californica	X				drainage
Melica frutescens	X				
Melica harfordii			X	X	cool; some summer water

Plant	Sun Dry	Sun Moist	Shade Dry	Shade Moist	Comments
Melica imperfecta			X	X	cool; some summer water
Melica torreyana			XX	X	drainage
Muhlenbergia montana		X			excellent drainage; higher elevations
Muhlenbergia richardsonis		X			excellent drainage; higher elevations
Muhlenbergia rigens		X			drought when established
Nassella cernua	X				drainage
Nassella lepida	X		X		drainage
Nassella pulchra	X				drainage
Nolina bigelovii	X				excellent drainage
Nolina parryi	X				excellent drainage
Odontostomum hartwegii	X				clay soils, serpentine
Panicum acuminatum		X			moist places
Phragmites australis		X			heat; inland; marshy sites
Pleuraphis jamesii	X				drainage; some summer water
Pleuraphis rigida	X				drainage
Poa secunda ssp. *secunda*	X				
Ptilagrostis kingii		X			drainage; high elevations
Schoenoplectus acutus var. *occidentalis*		X		X	wet places

Plant	Sun Dry	Sun Moist	Shade Dry	Shade Moist	Comments
Schoenoplectus tabernaemontani		X			wet places; invasive
Scirpus microcarpus		X			wet places
Scirpus robustus		X		X	wet places; coastal
Scoliopus bigelovii				X	coastal
Sisyrinchium bellum	X	X			summer dryness okay
Sisyrinchium californicum		X			year-round moisture; coastal
Sisyrinchium douglasii var. *douglasii*		X			drainage; not as easy as other species; higher elevations
Sisyrinchium elmeri		X			drainage; higher elevations
Sisyrinchium idahoense		X			excellent drainage; year-round moisture; higher elevations
Smilacina racemosa			X	XX	dry shade near coast okay
Smilacina stellata			X	XX	dry shade near coast okay
Smilax californica		X		X	
Sparganium emersum ssp. *emersum*		X		X	wet places
Sparganium eurycarpum ssp. *eurycarpum*		X		X	wet places
Sporobolus airoides		X			vernally moist
Stenanthium occidentale				X	drainage; wet places
Streptopus amplexifolius var. *americanus*				X	drainage; wet places

Plant	Sun Dry	Sun Moist	Shade Dry	Shade Moist	Comments
Trillium albidum				X	drainage; some summer moisture
Trillium angustipetalum				X	drainage; some summer moisture
Trillium chloropetalum				X	drainage; some summer moisture
Trillium ovatum				X	drainage; some summer moisture; coastal
Trillium rivale				X	drainage
Triteleia bridgesii	X				drainage
Triteleia hyacinthina	X				drainage
Triteleia ixioides		X		X	drainage; higher elevations
Triteleia ixioides ssp. *anilina*		X		X	excellent drainage; higher elevations
Triteleia ixioides ssp. *scabra*		X		X	some dryness okay
Triteleia laxa	X				drainage
Triteleia lilacina	X				drainage
Triteleia peduncularis	X	X			vernally moist
Typha angustifolia		X			wet places
Typha domingensis		X		X	wet places
Typha latifolia		X			wet places
Veratrum fimbriatum				X	lower elevations okay
Veratrum californicum var. *californicum*				X	higher elevations
Washingtonia filifera		X		X	deep, infrequent watering

Plant	Sun Dry	Sun Moist	Shade Dry	Shade Moist	Comments
Xerophyllum tenax	X				higher elevation; drainage; cool; acid soils
Yucca baccata	X				drainage
Yucca brevifolia	X				drainage
Yucca schidigera	X				drainage
Yucca whipplei	X				drainage
Zigadenus fremontii	X				drainage

Calochortus albus and *Dichelostemma* sp. (MARLIN HARMS)

SELECTED READINGS

Case, Frederick W., Jr., and Roberta B Case. *Trilliums*. Portland, OR: Timber Press, 1997.

Cohen, Victor A. *A Guide to the Pacific Coast Irises*. British Iris Society, 1967.

Connelly, Kevin. *Gardeners' Guide to California Wildflowers*. Sun Valley, CA: Theodore Payne Foundation, 1991.

Crampton, Beecher. *Grasses in California*. Berkeley: University of California Press, 1974.

Darke, Rick. *The Color Encyclopedia of Ornamental Grasses*. Portland, OR: Timber Press, 1999.

Emery, Dara E. *Seed Propagation of Native California Plants*. Santa Barbara, CA: Santa Barbara Botanic Garden, 1988.

Hickman, James C. (ed.). *The Jepson Manual: Higher Plants of California*. Berkeley: University of California Press, 1993.

Jacobs, Don L., and Rob L. Jacobs. *Trilliums in Woodland and Garden: American Treasures*. Decatur, GA: Eco-Gardens, 1997.

Jefferson-Brown, Michael and Kevin Pratt. *The Gardener's Guide to Growing Fritillaries*. Portland, OR: Timber Press, 1997.

Keator, Glenn. *Complete Garden Guide to the Native Perennials of California*. San Francisco: Chronicle Books, 1990.

Mathew, Brian. *Dwarf Bulbs*. New York: Arco Publishing Company, 1973.

North American Rock Garden Society. *Bulbs of North America*. Ed. Jane McGary. Portland, OR: Timber Press, 2001.

Phillips, Roger, and Martyn Rix. *The Bulb Book*. London: Pan Books, 1981.

Schmidt, Marjorie G. *Growing California Native Plants*. Berkeley: University of California Press, 1980.

Wilder, Louise Beebe. *Adventures with Hardy Bulbs*. New York, 1936; reprinted 1998 by Globe Pequot Press, Guilford, CT.

Eschscholzia californica and iris. (SAXON HOLT)

DISPLAY GARDENS

DESCANSO GARDENS
1418 Descanso Drive
La Cañada, CA 91011
(818) 952-4400
www.descanso.com

EAST BAY REGIONAL PARKS
    BOTANIC GARDEN
Wildcat Canyon Road and South Park
    Drive, Tilden Park
Berkeley, CA 94708
(510) 841-8732
www.ebparks.org
open daily, no admission fee
native plant display garden
plant sale third Saturday in April

LESTER ROWNTREE NATIVE
    PLANT GARDEN
(part of Mission Trail Park)
25800 Hatton Road, Carmel, CA
(408) 624-3543

MENDOCINO COAST
    BOTANIC GARDENS
18220 North Highway One
Fort Bragg, CA 95437
www.gardenbythesea.org

RANCHO SANTA ANA BOTANIC
    GARDEN
1500 North College Avenue
Claremont, CA 91711
(909) 625-8767
www.rsabg.org
open daily except major holidays
reference library, gift shop
native plants
plant sale first weekend in November

RUTH BANCROFT GARDEN
1500 Bancroft Road
PO Box 30845
Walnut Creek, CA 94598
(925) 210-9663
www.ruthbancroftgarden.org
succulent and xerophytic plants
plant sales, tours, gift shop

SANTA BARBARA BOTANIC
    GARDEN
1212 Mission Canyon Road
Santa Barbara, CA 93105
(805) 682-4726
www.sbbg.org
open daily, admission fee
gift shop, docent-led tours
native plants
spring and fall plant sales

STRYBING ARBORETUM AND
    BOTANIC GARDENS
Ninth Avenue and Lincoln Way,
    Golden Gate Park
San Francisco, CA 94122
(415) 661-1316
www.strybing.org
open daily
large native plant section
bookstore, horticulture library
plant sale first Saturday in May

UNIVERSITY OF CALIFORNIA
   BOTANICAL GARDEN
Centennial Drive
Berkeley, CA 94720
(510) 642-3343
www.mip.berkeley.edu/garden
open daily except Christmas, admission
   fee, parking fee
large native plant section
visitors' center/store

UNIVERSITY OF CALIFORNIA,
   DAVIS, ARBORETUM
La Rue Road
Davis, CA 95616
(916) 752-2498 (Friends of the Davis
   Arboretum)
www.arboretum.ucdavis.edu
open daily, no admission fee, parking fee
native plant display garden
plant sale in fall

RESOURCES

ALPINE GARDEN SOCIETY
AGS Centre, Avon Bank
Pershore
Worcestershire WR10 3JP
United Kingdom
(UK) 01386 554790
www.alpinegardensociety.org
seed exchange

AMERICAN HORTICULTURAL
    SOCIETY
7931 East Boulevard Drive
Alexandria, VA 22308
(703) 768-5700
bimonthly publication
seed exchange
gardeners information service
special events

AMERICAN IRIS SOCIETY
8426 Vinevalley
Sun Valley, CA 91352
(818) 767-5512
25 local affiliates in California

CALIFORNIA HORTICULTURAL
    SOCIETY
c/o Mrs. Elsie Mueller
1847 34th Avenue
Box WWW
San Francisco, CA 94122
(800) 884-0009 or (415) 566-5222
www.calhortsociety.org
monthly lectures, annual plant sale,
    seed exchange, field trips

CALIFORNIA NATIVE PLANT
    SOCIETY
2707 K Street, Suite 1
Sacramento, CA 95816
(916) 447-2677
www.cnps.org
30 chapters statewide with local
    newsletters, meetings, field trips,
    plant sales
state *Bulletin* and quarterly journal
    Fremontia

CALIFORNIA NATIVE GRASS
    ASSOCIATION
PO Box 72405
Davis, CA 95617
(916) 759-8458
newsletter, conferences, display
    gardens

INTERNATIONAL BULB SOCIETY
PO Box 336
Sanger, CA 93657
www.bulbsociety.com
annual and quarterly publications,
    seed exchange, web-based forums

NORTH AMERICAN LILY SOCIETY
c/o Dr. Robert Gilman, NALS
    Executive Secretary
PO Box 272
Owatonna, MN 55060
www.lilies.org
quarterly and annual publications,
    seed exchange

NORTH AMERICAN NATIVE
    ORCHID ALLIANCE
PO Box 772121
Ocala, FL 34477
(352) 861-2565 or (207) 636-3719
www.flmnh.ufl.edu/naorchid
annual conference, quarterly journal

NORTH AMERICAN ROCK GAR-
    DEN SOCIETY
PO Box 67
Millwood, NJ 10546
quarterly bulletin, seed exchange

PACIFIC BULB SOCIETY
c/o Vicki Sironen
PO Box 906
Preston, WA 98050
affiliate of International Bulb Society

PACIFIC HORTICULTURE
PO Box 680
Berkeley, CA 94701
(510) 849-1627
quarterly journal

SOCIETY FOR PACIFIC COAST
    NATIVE IRIS
c/o Richard C. Richards
5885 Cowles Mt. Blvd.
La Mesa, CA 91942
www.pacificcoastiris.org
newsletter, specialized publications,
    checklists

SPECIES IRIS GROUP OF NORTH
    AMERICA (SIGNA)
c/o Rodney Barton
3 Wolters Street
Hickory Creek, TX 75065
www.signa.org
newsletter, seed exchange

SPECIES LILY PRESERVATION
    GROUP
4234 Randhurst Way
Fair Oaks, CA 95628
www.lilies.org/slpg
annual meeting, newsletter

NURSERIES AND SEED SUPPLIERS

ALBRIGHT SEED COMPANY
(800) 423-8112
www.albrightseed.com
retail with minimum order at website

ANDERSON VALLEY NURSERY
18151 Mountain View Road
PO Box 504
Boonville, CA 95415
(707) 895-3853
Fax: (707) 895-2850
retail/wholesale/no mail order
weekends and by appointment
call for retail availability

BAY VIEW GARDENS
1201 Bay Street
Santa Cruz, CA 95060
(831) 423-3656
Fax: (831) 423-7610
email: ghrobayview@surfnetusa.com
mail order/retail
call for catalog

BERKELEY HORTICULTURAL
    NURSERY
1310 McGee Avenue
Berkeley, CA 94703
(510) 526-4704
retail/no mail order
call for hours

B & T WORLD SEEDS
Route des Marchandes
Paguignan, 34210 Olonzac
France
phone: 33 04 6891 29 63
www.b-and-t-world-seeds.com

CALIFORNIA FLORA NURSERY
2990 Somers Street
PO Box 3
Fulton, CA 95439
(707) 528-8813
retail/wholesale
call for hours and plant list

C. H. BACCUS
900 Boynton Avenue
San Jose, CA 95117
(408) 244-2923
retail/wholesale
by appointment only

CORNFLOWER FARMS
PO Box 896
Elk Grove, CA 95759
(916) 689-1015
Fax: (916) 689-1968
open to public 2nd Sat, Apr–Nov,
    7am–1pm
www.cornflowerfarms.com
wholesale/mail order
call for catalog

ELKHORN NATIVE PLANT
    NURSERY
1957B Highway 1
PO Box 270
Moss Landing, CA 95039
(831) 763-1207
Fax: (831) 763-1659
email: enpn@elkhornnursery.com
www.elkhornnursery.com
retail/wholesale/mail order
wholesale Mon–Fri 8am–4pm; retail
    Wed 10am–3pm

FAR WEST BULB FARM
1449 Lower Colfax Road
Grass Valley, CA 95945
(530) 272-4775
email: nancyames@accessbee.com
website:
    www.californianativebulbs.com
retail/mail order only

FLORAL NATIVE NURSERY
2511 Floral Avenue
Chico, CA 95973
(530) 892-2511
Fax: (530) 342-1641
retail
call for plant list

FRASER'S THIMBLE FARMS
175 Arbutus Road
Salt Spring Island, B.C.
Canada V8K 1A3
(205) 537-5788
Fax: 250-537-5788
www.thimblefarms.com
retail/mail order/no phone orders
Mar–Jun 10am–4:30pm daily; Jul–Feb
    10am–4:30pm Tue–Sun
catalog and order form at website

FRESHWATER FARMS
5851 Myrtle Avenue
Eureka, CA 95503
(707) 444-8261 or (800) 200-8969
Fax: (707) 442-2490
email: info@freshwaterfarms.com
www.freshwaterfarms.com
retail/wholesale/mail order
closed Sun
catalog and seed list at website

GUALALA NURSERY & TRADING
    COMPANY
38700 South Highway One
PO Box 957
Gualala, CA 95445
(707) 884-9633
retail/wholesale
7 days a week
call for plant list

HEDGEROW FARMS (grasses)
21740 County Road 88
Winters, CA 95694
(530) 662-4570
Fax: (530) 668-8369
www.hedgerowfarms.com
retail/wholesale/mail order
call for catalog

HERONSWOOD NURSERY
7530 N.E. 288th Street
Kingston, WA 98346
(360) 297-4172
Fax: (360) 297-8321
email: info@heronswood.com
www.heronswood.com
mail order/retail/open by appointment
    only

INTERMOUNTAIN NURSERY
30443 North Auberry Road
Prather, CA 93651
(559) 855-3113
retail/wholesale/no mail order
plant list with SASE
Mon–Sat 8am–5pm, Sun 10am–4pm

JEFF ANHORN NURSERY
PO Box 2061
Livermore, CA 94551
(925) 447-0858
Fax: (925) 447-1854
wholesale/no mail order
by appointment only
plant list with SASE

LARNER SEEDS
PO Box 407
Bolinas, CA 94924
(415) 868-9407
www.larnerseeds.com
retail/wholesale/mail order
call for hours and catalog

LAS PILITAS NURSERY
3232 Las Pilitas Road
Santa Margarita, CA 93453
(805) 438-5992
retail/wholesale/mail order
www.laspilitas.com
retail Fri–Sat 9am–5pm; wholesale
    Mon–Fri 8am–6pm
catalog at website

LAS PILITAS NURSERY SOUTH
8331 Nelson Way
Escondido, CA 92026
(760) 749-5930
www.laspilitas.com
retail/wholesale
Mon–Sat 9am–4pm
catalog at website

LOS OSOS VALLEY NURSERY
301 Los Osos Valley Road
Los Osos, CA 93402
(805) 528-5300
retail
daily except Christmas and
    New Year's Day
call for plant list

MATILIJA NURSERY
8225 Waters Road
PO Box 429
Moorpark, CA 93021
(805) 523-8604
email: matilija@verizon.net
www.matilijanursery.com
wholesale/retail
Mon–Thu 8:30am–noon, Fri–Sat
    8:30am–2pm
retail by appointment
call for directions, plant list

MENZIES NATIVES NURSERY
10805 N. Old Stage Road
PO Box 9
Weed, CA 96094
(530) 938-4858
Fax: (530) 938-4777
retail/wholesale/mail order
plant list with SASE
call for hours

MOSTLY NATIVES NURSERY
27235 Highway One
PO Box 258
Tomales, CA 94971
(707) 878-2009
Fax: (707) 878-2079
www.mostlynatives.com
retail/no mail order
hours and plant list at website

NATIVE HERE NURSERY
101 Golf Course Drive
Berkeley, CA 94708
(510) 549-0211
www.ebcnps.org
retail/wholesale
call for hours

NATIVE REVIVAL NURSERY
2600 Mar Vista Drive
Aptos, CA 95003
(831) 684-1811
email: plants@nativerevival.com
www.nativerevival.com
Tue–Fri 8:30am–5pm, Sat–Sun
    10am–4pm
retail/wholesale
call for hours and catalog

NATIVE SONS WHOLESALE
    NURSERY
379 West El Campo Road
Arroyo Grande, CA 93420
(805) 481-5996
Fax: (805) 489-1991
email: native.son@nativeson.com
www.nativeson.com
wholesale only
Mon–Fri 8am–4:30pm, Sat 8am–noon
catalog @ website

NORTH COAST NATIVE
    NURSERY
2710 Chileno Valley Road
PO Box 744
Petaluma, CA 94953
(707) 769-1213
www.northcoastnativenursery.com
wholesale Mon–Fri 8am–4pm
retail by appointment only
call for plant list or see website

PACIFIC RIM NATIVE PLANT
    NURSERY
44305 Old Orchard Road
Chilliwack, B.C.
Canada V2R 1A9
(604) 792-9279
Fax: (604) 792-1891
email: plants@hillkeep.ca
www.hillkeep.ca
retail/wholesale/mail order
catalog at website

PAUL CHRISTIAN RARE PLANTS
PO Box 468
Wrexham, LL13 9XR, United Kingdom
Phone: 44 01978 366399
www.rareplants.co.uk

PLANET EARTH GROWERS
    NURSERY
6931 East Belmont Ave
Fresno, CA 93727
(559) 251-5344
Fax: (559) 251-5539
www.planetearthnursery.com
retail/wholesale/no mail order
call for hours and plant list

REVEG EDGE SERVICES
PO Box 609
Redwood City, CA 94064
(650) 325-7333
wholesale/mail order
all plants custom grown from customer-
    supplied seeds
catalog at www.ecoseeds.com

RED TAIL FARMS
9000 Busch Lane
Potter Valley, CA 95469
(707) 743-2734
wholesale only
call for plant list

REDWOOD CITY SEED COMPANY
PO Box 361
Redwood City, CA 94064
(650) 325-7333
www.ecoseeds.com
wholesale/mail order
all plants custom grown from customer-
supplied seeds

ROSENDALE NURSERY (wholesale
only)
Sierra Azul (retail nursery)
2660 E. Lake Ave
Watsonville, CA 95076
(831) 728-2599
Fax: (831) 728-2537
email:
sierrarosendale@compuserve.com
www.sierraazul.com
9am–4:30pm daily

S AND S SEEDS
PO Box 1275
Carpinteria, CA 93014
(805) 684-0436
Fax: (805) 684-2798
email: infor@ssseeds.com
www.ssseeds.com
wholesale/mail, fax, and phone order
only
call for catalog

SIERRA SEED SUPPLY
PO Box 42
Greenville, CA 95947
(530) 284-7926
mail order only/retail
call for seed catalog

SISKIYOU RARE PLANT NURSERY
2525 Cummings Road
Medford, OR 97501
(541) 772-6846
Fax: (541) 772-4917
email: customerservice@srpn.net
www.srpn.net
retail/wholesale/mail order/online order
catalog at website

SUNCREST NURSERIES
wholesale only
www.suncrestnurseries.com
list of retail nurseries at website

TAHOE TREE COMPANY
401 West Lake Blvd
Tahoe City, CA 96145
(916) 583-3911
retail/wholesale/no mail order
open year round

TELOS RARE BULBS
PO Box 4147
Arcata, CA 95518
email: rarebulbs@earthlink.net
www.telosrarebulbs.com
mail order/retail
write or email for catalog

THEODORE PAYNE
FOUNDATION
10459 Tuxford Street
Sun Valley, CA 91352
(818) 768-1802
www.theodorepayne.org
retail
call for hours and catalog

TREE OF LIFE NURSERY
33201 Ortega Highway
PO Box 635
San Juan Capistrano, CA 92693
(949) 728-0685
Fax: (949) 728-0509
www.treeoflifenursery.com
wholesale Mon–Fri 8am–4:30; retail Fri
    9am–4pm
call for catalog

VILLAGER NURSERY
10678 Donner Pass Road
Truckee, CA 96161
(530) 587-0771
retail/wholesale/no mail order
call for hours
plant list with SASE

WAYWARD GARDENS
1296 Tilton Road
Sebastopol, CA 95472
(707) 829-8225
www.waywardgardens.com
retail/no mail order
call for plant list

YERBA BUENA NATIVE PLANT
    NURSERY
19500 Skyline Blvd
Woodside, CA 94062
(650) 851-1668
Fax: (650) 851-5565
www.yerbabuenanursery.com
retail
Tue–Sun 9am–5pm
catalog at website

PHOTOGRAPHERS

Carol Bornstein
Santa Barbara Botanic Garden
1212 Mission Canyon Road
Santa Barbara, CA 93105

Robert Case
2945 Corte Miguel
Concord, CA 94518

Fran Cox
3201 Plumas Street, #130
Reno, NV 89509

Steve Edwards
East Bay Regional Parks Botanic Garden
Tilden Regional Park
Berkeley, CA 94708

William Follette
1 Harrison Avenue
Sausalito, CA 94965

Marlin Harms
563 Estero Avenue
Morro Bay, CA 93442

Saxon Holt
31 Bay Tree Hollow
Novato, CA 94945

Stephen Ingram
140 Willow Road,
Swall Meadows
Bishop, CA

Steve Junak
Santa Barbara Botanic Garden
1212 Mission Canyon Road
Santa Barbara, CA 93105

Charles Kennard
76 Suffield Avenue
San Anselmo, CA 94960

Bart O'Brien
Rancho Santa Ana Botanic Garden
1500 N. College Avenue
Claremont, CA 91711

Roger Raiche
12440 Occidental Road
Sebastopol, CA 95472

Doreen Smith
252 Mt. Shasta Drive
San Rafael, CA 94903

Robert C. West
42 Diamond Head Passage
Corte Madera, CA 94925

Erythronium californicum 'White Beauty'. (SAXON HOLT)

INDEX